The Internet and Philosophy of Science

'This book analyzes the Internet from the point of view of philosophy of science, thereby providing new insights into the key technology of our time.'

Donald Gillies, *University College London, UK*

From the perspective of the philosophy of science, this book analyzes the Internet conceived in a broad sense. It includes three layers that require philosophical attention: (1) the technological infrastructure, (2) the Web, and (3) cloud computing, along with apps and mobile Internet. The study focuses on the network of networks from the viewpoint of complexity, both structural and dynamic. In addition to the scientific side, this volume considers the technological facet and the social dimension of the Internet as a novel design.

There is a clear contribution of the Internet to science: first, the very development of the network of networks requires the creation of new science; second, the Internet empowers scientific disciplines, such as communication sciences; and third, the Internet has fostered a whole new emergent field of data and information. After the opening chapter, which offers a series of keys to the book, there are nine chapters, grouped into four parts: (I) Configuration of the Internet and Its Future, (II) Structural and Dynamic Complexity in the Design of the Internet, (III) Internal and External Contributions of the Internet, and (IV) The Internet and the Sciences.

Following this framework, *The Internet and Philosophy of Science* will be of interest to scholars and advanced students working in philosophy of science, philosophy of technology as well as science and technology studies.

Wenceslao J. Gonzalez is Professor of Logic and Philosophy of Science (University of A Coruña). He is a Full Member of the *Académie Internationale de Philosophie des Sciences*/International Academy for Philosophy of Sciences. He was a Team Leader of the European Science Foundation program entitled 'The Philosophy of Science in a European Perspective' (2008–2013). He has been *Visiting Fellow* at the Center for

Philosophy of Science (University of Pittsburgh) and a visiting researcher at the London School of Economics.

Gonzalez has been a member of the National Committee for Evaluation of the Scientific Activity (CNEAI) of Spain. He is the director of the Center for Research in Philosophy of Science and Technology (CIFCYT) at the University of A Coruña, where he organizes an annual conference on Contemporary Philosophy and Methodology of Science. His publications include monographs such as *Philosophico-Methodological Analysis of Prediction and its Role in Economics* (2015) and the edition of 44 volumes on philosophy of science and technology, such as *New Approaches to Scientific Realism* (2020). He is the author of numerous papers on the Internet from a philosophical perspective, such as 'The Internet at the Service of Society: Business Ethics, Rationality, and Responsibility' (2020).

Routledge Studies in the Philosophy of Science

For more information about this series, please visit: https://www.routledge.com/Routledge-Studies-in-the-Philosophy-of-Science/book-series/POS

The Internet and Philosophy of Science

Edited by Wenceslao J. Gonzalez

Routledge
Taylor & Francis Group

NEW YORK AND LONDON

First published 2023
by Routledge
605 Third Avenue, New York, NY 10158

and by Routledge
4 Park Square, Milton Park, Abingdon, Oxon, OX14 4RN

*Routledge is an imprint of the Taylor & Francis Group,
an informa business*

© 2023 selection and editorial matter, Wenceslao J. Gonzalez;
individual chapters, the contributors

Library of Congress Cataloging-in-Publication Data
A catalog record for this title has been requested

ISBN: 978-1-032-16457-1 (hbk)
ISBN: 978-1-032-16832-6 (pbk)
ISBN: 978-1-003-25047-0 (ebk)

DOI: 10.4324/9781003250470

Typeset in Sabon
by SPi Technologies India Pvt Ltd (Straive)

Contents

Contributors

Enrique Alonso (Autonomous University of Madrid), Associate Professor of Logic and Philosophy of Science. He is the author of the book *Contenido, verdad y consecuencia lógica* [Content, truth and logical consequence] (1995) and co-editor of *Los viajes en el tiempo: Un enfoque multidisciplinar* [Time Travel: A Multidisciplinary Approach] (2009) and *Una perspectiva de la Inteligencia Artificial en su 50 aniversario* [A Perspective on Artificial Intelligence on its 50th Anniversary] (2006). He has also published 'De la computabilidad a la hipercomputación' [From computability to hypercomputing] (2006), 'El debate público en las redes sociales: *Twitter*. España como estudio de caso' [The public debate on social networks: Twitter. Spain as a case study] (2013) and *El nuevo Leviatán. Una Historia Política de la Red* [The new Leviathan. A Political History of the Network] (2015).

Maria Jose Arrojo is Associate Professor at the University of A Coruña. She has been a member of the Research Group of Philosophy and Methodology of Artificial Sciences of the University of A Coruña since its constitution and has been working on these issues for 17 years. Dr Arrojo is also a member of the Center for Research in Philosophy of Science and Technology (CIFCYT) and a board member of the Foundation Philosophy of Science and Technology [*Fundación Filosofía de la Ciencia y la Tecnología*].

Arrojo has multiple publications on philosophy and methodology of communication sciences, focused from the perspective of the design sciences. Among them, two are with Gonzalez: 'Diversity in Complexity in Communication Sciences: Epistemological and Ontological Analyses' (2015), and 'Complexity in the Sciences of the Internet and its Relation to Communication Sciences' (2019).

Wenceslao J. Gonzalez is Professor of Logic and Philosophy of Science (University of A Coruña). He is a Full Member of the *Académie International de Philosophie des Sciences*/International Academy for Philosophy of Sciences. He has been a Team Leader of the European Science Foundation program entitled 'The Philosophy of Science in a

European Perspective' (2008–2013). He has been *Visiting Fellow* at the Center for Philosophy of Science (U. Pittsburgh) and a visiting researcher at the London School of Economics.

Gonzalez has been a member of the National Committee for Evaluation of the Scientific Activity (CNEAI) of Spain. He is the director of the Center for Research in Philosophy of Science and Technology (CIFCYT) at the University of A Coruña, where he organizes every year a conference on Contemporary Philosophy and Methodology of Science. His publications include monographs such as *Philosophico-Methodological Analysis of Prediction and its Role in Economics* (2015) and the edition of 44 volumes on philosophy of science and technology, such as *New Approaches to Scientific Realism* (2020). He is the author of numerous papers on the Internet from a philosophical perspective, such as 'The Internet at the Service of Society: Business Ethics, Rationality, and Responsibility' (2020).

James A. Hendler is Director of the *Institute for Data Exploration and Applications* and *Tetherless World Professor of Computer, Web and Cognitive Sciences* at *Rensselaer Polytechnic Institute* (Troy, New York). He is one of the proponents of the semantic Web and author of 350 scientific papers (including books, articles and technical papers). Hendler's first degree was from Yale University (1978). He completed two Master's degrees (1981 and 1983). His Ph.D. was in Philosophy at Brown University (1986). From that year until 2006, he worked at the University of Maryland, where he was Professor (1999–2006) and director of the *Joint Institute for Knowledge Discovery*.

Hendler is a Fellow of the American Association for Artificial Intelligence, the British Computer Society, the Institute of Electrical and Electronics Engineers and the American Association for the Advancement of Science (AAAS). He has been principal investigator on major projects funded by several U.S. national agencies. Among his book is *Social Machines: The Coming Collision of Artificial Intelligence, Social Networks and Humanity* (2016).

Ole Hanseth is Professor of Digitalization and Entrepreneurship, Department of Computer Science (University of Oslo). He is co-editor of the books *From Control to Drift: The Dynamics of Corporate Information Infrastructures* (2000), *Risk, Complexity and ICT* (2007), and *Information Infrastructures within European Health Care: Working with the Installed Base* (2017). He is co-author of the article 'Design Theory for Dynamic Complexity in Information Infrastructures: The Case of Building Internet' (2010) and also of the paper 'Infrastructural Innovation: Flexibility, Generativity and the Mobile Internet' (2013).

Yan Luo is Professor of the Department of Electrical and Computer Engineering (University of Massachusetts Lowell, Boston). He obtained his Ph.D. in Computer Science from University of California Riverside in 2005. His research interest spans machine learning based on big data analytics, network design, heterogeneous computing and embedded sensor systems. He has served on the editorial board of several peer-reviewed journals and as a program chair of several IEEE conferences. His research contributions on big data, network processor design and programmable networking have led to his view on the design philosophy of the future Internet.

Thanassis Tiropanis is Associate Professor at the University of Southampton, member of the *Web and Internet Science Group, Electronics and Computer Science*. He did his doctoral thesis in *Computer Science* at University College London and holds a degree in *Computer Engineering and Informatics* from the University of Patras (Greece). He is co-author of the seminal paper 'Network Science, Web Science, and Internet Science' (2015) and has collaborated in the elaboration of the widely cited report on *European Digital Infrastructure and Data Sovereignty: A Policy Perspective* (EIT, European Union, Brussels, 9.6.2020).

Ruth Towse has been Professor of Economics of Creative Industries at Bournemouth University, UK, since 2007, where she is Co-Director for Economics at CIPPM (the Centre for Intellectual Property Policy and Management) and CREATe Fellow in Cultural Economics (University of Glasgow). She previously held a number of university positions in the UK, Thailand and at Erasmus University Rotterdam (1999–2008). She specializes in cultural economics and the economics of copyright. She has published widely on both fields in academic journals and books, including *A Textbook of Cultural Economics* (Cambridge University Press), now in its 2nd edition, and the *Handbook of Cultural Economics*, now in its 3rd edition (with Trilce Navarrete Hernandez). She is a co-editor of the book *The Internet and the Mass Media* (2008).

Towse was Joint Editor of the *Journal of Cultural Economics* from 1993–2002, President of the Society for Economic Research in Copyright Issues from 2004–2006 and President of the Association for Cultural Economic International (ACEI) from 2006–2008. In 2016, she was made Distinguished Fellow of the ACEI.

1 Creativity in the Internet as a Complex Setting

Interaction between Scientific Creativity, Technological Innovation and Social Innovation[1]

Wenceslao J. Gonzalez

1.1 A Perspective of Creativity in a Setting of Complexity

Two angles of the Internet – in the broad sense – have often been dominant in philosophico-methodological analysis: the technological and the social.[2] This is reflected in the importance given to technological innovation, considered in itself and in its economic repercussion, and in social innovation, seeing the impact of the network of networks in the knowledge society.[3] The technological and social angles are emphasized when Joseph Schumpeter's ideas on invention, innovation and diffusion are used in this realm to maintain that 'the original invention is followed by subsequent innovation, and even in the diffusion phases, new developments will materialise' (Henten and Tadayoni 2008, p. 47).

But the scientific angle is also important, because of the multivariate relationship between the Internet – in the broad sense – and science. Furthermore, the scientific side interacts with the technological facet and the social dimension within the complex setting of the network of networks. This analysis is carried out here from the perspective of *Internet creativity*. It is a creativity with a triple origin – scientific, technological and social – which presupposes at least three elements: (1) the existence of an agent (individual or social) that initiates a process that introduces variation with respect to what is known; (2) a directly sought end, related to one or more of the spheres mentioned; and (3) a selected way of doing things, which is appropriate to obtain the sought variation.

This creativity is developed within the setting of the Internet as a complex system, which has to face problems that may not be well defined and require a novel and appropriate solution[4] (among these problems are issues related to fake news). The kind of complexity involved in the Internet in the broad sense is structural and dynamic, but also pragmatic. Thus, complexity affects the epistemological and ontological configuration of the network of networks, its change over time, as well as the diversification due to different contexts of use and increasing personalization (especially on the Web and with apps, certain uses of the cloud computing and some expressions of the mobile Internet).[5]

DOI: 10.4324/9781003250470-1

When the type of creativity is *scientific*, then it is different from the creativity of content that, with the support of the network, is conveyed through the Web or using the mobile Internet.[6] In this regard, there are three main philosophico-methodological components of the network of networks: (a) the scientific side, (b) the technological facet, and (c) the social dimension.[7] These three components intervene in the artificial reality of Internet in the broad sense, which arises from a layered design.[8] This ontology of artificial strata includes structural and dynamic factors – and also pragmatic traits – and serves as a basis for operational issues of this complex system, including economic ones.[9] Furthermore, the layers are relevant for its governance, where centralization and decentralization play an important role in the present and future of this network of networks.

Three are the large layers or ontological strata of this network of networks:[10] (i) the Internet as a technological infrastructure that serves as a common or 'inter-network' support for this complex information and communication system; (ii) the Web as a shared space for communication of people (webmail, social networks, YouTube, webpages, etc.), instead of a mere operational interweaving of machines; and (iii) cloud computing, along with the practical applications (apps), which expand the potential of the human communication to reach new social possibilities, and the 'mobile Internet,'[11] which makes the contents almost ubiquitously accessible and promotes a pragmatic complexity.

For the philosophico-methodological study of *creativity* in the Internet,[12] several steps are followed here: (1) to highlight that the structural and dynamic *interaction* of scientific creativity, technological innovation and social innovation is key to the creativity of the network of networks; (2) the recognition that interaction occurs upon an *ontological basis*, consisting of three large layers or ontological strata of this network of networks; (3) creativity in science in the realm of the Internet – conceived in the broad sense – is *multivariate*, as it is diversified into three groups of scientific disciplines, which reinforces the idea of an epistemological and methodological pluralism in this artificial sphere;[13] (4) each technological innovation – mainly epistemological, methodological and ontological – is commonly closely connected with scientific creativity,[14] especially in the case of the network; (5) social innovation shows greater creativity in the layers of the Web and the practical applications (apps) and 'mobile Internet' than in the network;[15] and (6) creativity is needed regarding communication through the network of networks to deal with relevant problems, such as fake news.

1.2 Scientific Creativity, Technological Innovation and Social Innovation

Overall, there are three leading factors for creativity on the Internet, conceived in a broad sense,[16] which is the complex system where the network, the Web and the cloud computing with the practical applications (apps) and mobile Internet are together in a structural and dynamic interaction. The main *factors for creativity* of the network of networks are scientific creativity, technological innovation and social innovation. This is the case because of the three central philosophico-methodological components of the Internet, which are also the main sources of complexity of this system: the scientific side,[17] the technological facet and the social dimension.[18] They are intertwined in many ways in the network of networks, rather than being in a linear relationship or in a unidirectional path.

First, *scientific creativity* develops its role regarding the Internet – in the broad sense – in several directions: (i) as a central aspect of the sciences of the artificial that, through new designs, seeks novel aims, processes and results to expand the human possibilities generated by the use of the network of networks;[19] (ii) as key support for new technological advances, mainly in the area of information and communication technology (ICT), since scientific creativity provides a key knowledge for technological innovation;[20] and (iii) as an integral part of a social activity that seeks to meet the demands of society (individuals, groups, organizations, governments, etc.), where the social agents are not mere recipients of scientific creativity,[21] because they can also play an active role in promoting their development by demanding new scientific aims, processes and results in the network of networks.

Creativity in science is multivariate, because it can be connected with each of its constituent elements.[22] Thus, we find creativity in language, structure, knowledge, method, activity, aims and values of science in general, of a group of sciences (natural, social or artificial) and of a particular science. This creativity in science affects this complex human activity, both in its structure and in its dynamics. This is what happens on the Internet as a complex system, which is aimed at each of its layers (i.e., the network, the Web and the cloud computing, apps and mobile Internet). Both the science use and the network of networks are open to improvements now and in the future. Meanwhile, innovation has a key role in technology,[23] insofar as it is conceived to carry out a *creative transformation* of the reality: the search for something new or better than an alternative developed by another business firm or public enterprise. Thus, innovation always has a role in technology, and it can be connected to the central components of the technology.[24]

Second, *technological innovation* has grown from the beginning of the network (that is, from the time of Arpanet, around the end of the 1960s and the beginning of the 1970s) to the present day.[25] The growth has been in two directions: horizontal or longitudinal and vertical or transversal. The first has occurred by expanding the possibilities of something existing (such as fiber optics to transmit information, electronic commerce,[26] digital radio or digital television) and the second has occurred through the creation of new technologies to achieve completely new objectives, giving support to them (such as email and webmail, social networks, blockchain for the cryptocurrencies like bitcoin,[27] cloud computing, the Internet of Things, etc.).

Nevertheless, even if it happens that technological change 'is not just inevitable but beneficial' (Clark 2018, p. 170), there are limits to Internet innovation as the network of networks.[28] Some limits are ontological (like physical limits), others are epistemological (frontiers of our knowledge), whereas others are practical or impairments, which can be either contextual or structural. In this regard, the network has proven to be able to overcome a number of impediments over time.[29] Thus, it has shown a dynamic, which is internal as well as external, and which can be explained in terms of historicity.[30]

On the one hand, technology platforms may be stable for a while, but both the network and its applications may be subject to *internal* pressure to vary via innovation, which may lead to its subsequent replacement.[31] This is reflected in the designs that engineers have to make, where how to approach the problem is the key to finding the right solution.[32] On the other hand, there may be *external* pressure to introduce changes due to variations in the requirements demanded by new contexts.[33] For David Clark, what causes network requirements to change over time can be explained as 'the interplay among three important drivers: new developments in network and computer technology, new approaches to application design, and changing requirements in the larger context in which Internet sits' (Clark 2018, p. 301). Obviously, the social dimension plays a crucial role in the component of the context, which directly affects the external dynamics of the network of networks.

Third, *social innovation* with respect to the network of networks can be seen at three levels: micro (contribution of individuals or groups), meso (involvement of business firms or organizations) and macro (changes due to big corporations – Alphabet [Google], Facebook, Apple, Amazon, Microsoft, Huawei, Alibaba, etc. – or by governments or international organizations). Commonly, the technological giants have a combination of several aspects: software (based on scientific creativity), hardware (developed by technological innovation) and services (of interest for the social dimension).[34]

Given the complex system of the Internet (the network, the Web and the cloud computing with the practical applications [apps]) as a whole, there has been a powerful social innovation that has contributed to the transition of the Internet from its public origins as a technological platform focused on information – with pluralistic tendencies – to an increasingly commercial network of networks.[35] Social innovation is not a mere exterior influence, insofar as the new social demands on the Internet – in a broad sense – can come from *inside* (scientists and technologists)[36] and from *outside* (the users) of the network of networks. Each of these groups can influence the selection of new objectives, different processes, novel results or more beneficial consequences, which are relevant *within* the complex system of the Internet but are also commonly important from *without* the network itself.

As the development of the Internet as a whole increasingly depends on factors that are external to the network of networks (social, economic, political, ecological, etc.), what was initially 'external' becomes internalized in scientific terms or from a technological point of view, in the search for new aims, processes, results and consequences. This explains the trajectory followed by the Web since its open access to the general public (April 30, 1993)[37] and what happened with the cloud and the constellation of practical applications (apps) currently available (e.g., in the 4G and 5G mobile phones).

1.3 Three Large Layers, or Ontological Strata, of This Network of Networks

Ontologically, there are three main strata in the network of networks, where the three expressions of complexity can be found: structural, dynamic and pragmatic. These strata are the layers of the technological infrastructure, which is the Internet *sensu stricto*; the Web with all its components (social networks, etc.); and cloud computing, along with the practical applications (apps) and mobile Internet. Scientific creativity, technological innovation and social innovation have a role (epistemological, methodological, operational, etc.) in these large artificial ontological strata of the Internet in the broad sense.

Each of these three layers of the network of networks shows an interaction between scientific creativity and technological innovation, which includes the social dimension in several ways.[38] The interaction science-technology is initially a bidirectional relationship, but it becomes multidirectional because of the presence of the social dimension. To properly grasp the bidirectional relationship between scientific creativity and technological innovation, we must start from a vision of *diversified scientific activity* in basic science, applied science and application of science.[39] At the same time, technological innovation is related to at least three key elements: knowledge (scientific, technological and evaluative), the task of transforming reality through actions and products or artifacts.[40]

Following this intertwined relationship of scientific creativity and technological innovation, advancement in each of the layers is possible, insofar as (I) scientific creativity and technological innovation are led by the search for structural or dynamic *originality* and (II) they seek some kind of *novelty* (semantic, epistemological, methodological, ontological, axiological, etc.) as a result of that search for something original in the Internet as a complex system, where there might be the emergence of new properties.[41] The consequence is that the situation in the network, the Web and the cloud seems to be methodologically twofold and also bidirectional, because it is the case that *creativity* can have a role in invention[42] and *innovation* can lead to creativity.[43]

Technological innovation comes primarily with the Internet in a strict sense – the infrastructure – which serves as the basis for the complex system of the network of networks. Above all, it does so with the two main types of technology that are used in this complex system: (a) the background or backbone technology for the network, followed by the direct support as service to the users and (b) the information and communication technologies (ICT). In the former, we can distinguish between an Internet Infrastructure Provider (IIP),[44] which is the bedrock of the complex system, and the Internet Service Provider (ISP), which is the company that gives users access to the Internet (a 'cable company' that actually runs the technology over the basic infrastructure and provides the allocated data of the user).[45]

Frequently, the technological focus is on the second element of the background technology for the network, which is directly related to the information and communication technologies of the users (individuals, groups, organizations, governments, etc.). Thus, it happens that users commonly focus on the available applications (and possible creation of new ones), where the pragmatic complexity is noticeable, within a complex system that is an artificial environment. Meanwhile, technology experts are especially focused on the support, i.e., the packet transport service provided by the Internet service providers, where the applications run *on top of* that service.[46]

Clearly, there is, in the network, a close relation to creativity, insofar as innovation involves some kind of change that goes *beyond tradition* in order to achieve specific goals, which are usually new in this case. Initially, there is always a chance that the innovation might be obtained by trial and error. But the evolution of innovation over the decades calls for a more systematic procedure, i.e., the use of principles and strategies,[47] where scientific creativity has a role. It seems clear that any innovation looks for something possible in the *future* that is expected on the basis of present knowledge and means.[48] When innovation is longitudinal, then it might be seen 'as a process of searching for better combinations of existing components' (Fleming and Sorenson 2004, p. 910). But when innovation is transversal, there is more novelty achieved and what we get can sometimes be considered as a 'technological revolution.'

Heuristically, there might be two main situations. The first is when inventors in technology can combine relatively independent components (i.e., those with a low degree of coupling) and they can find useful new configurations. In these cases, the process of innovation might be relatively easy (i.e., any search algorithm can locate the most useful combinations). The second is when the space searched becomes increasingly complex. This involves that local search routines break down. Consequently, they might fail to identify the best combinations.[49] In these cases, science is particularly useful: It 'may transform invention from a relatively haphazard search process to a more directed identification of useful new combinations, thus mitigating the complications typically encountered when combining coupled components' (Fleming and Sorenson 2004, p. 910).

An example of interdependence of creativity in the sphere of the sciences of the artificial and innovation in the realm of information and communication technologies can be found in the professional trajectory of Tim Berners-Lee, who started the second layer: the World Wide Web. He developed scientific creativity by laying the foundations of Web Science at a research center (CERN, *Conseil Européen pour la Recherche Nucléaire*) and brought about technological innovation through ICT using NeXT computers.[50]

According to James Hendler, Tim Berners-Lee proposed in 1989 a hypertext system that could be one of the applications that ran on top of the network (the Internet in the strict sense). 'WorldWideWeb' (originally one word) was the name used in 1990 for a formal proposal of the program, which was published and approved. This new program consisted of several components, which include three fundamental building blocks: (i) URL as a protocol for naming the items on the Web, (ii) HTTP as a protocol for how machines would find each other using the Internet and exchange information and (iii) HTML as a language for how the exchanged information would be displayed in a program called a 'browser.' *World Wide Web* was the name released in 1991 to put together these elements.[51]

As a matter of fact, the social dimension of the Internet as a complex system is often linked to the use of the Web, such as webmail, which actually belongs to the 'external side' of the network of networks, or other uses of the Web 2.0,[52] such as YouTube, which is another successful modern Web application that 'allows users to upload video content to the Web, and provides a number of mechanisms for letting users to share this content' (Hendler and Golbeck 2008, p. 16). In this regard, according to James Hendler and Jennifer Golbeck, 'the social nature of the Web 2.0 sites primarily allows linking between people, not content, thus creating large, and valuable, social networks, but with impoverished semantic value among the tagged content' (Hendler and Golbeck 2008, p. 15).

Once the Web had been developed, there was a new interplay between scientific creativity, technological innovation and social innovation that led to the creation of cloud computing and multiple practical applications (apps) as well as the mobile Internet. This third layer, or ontological stratum, of the network of networks allows the accumulation of data and information – the case of the cloud – which are useful for various scientific activities (especially for science data, which grasps emerging properties derived from the use of the Internet in a broad sense) and encourages the development of practical utilities for the users of the network of networks through the apps, which have a similar configuration – architecture – to the Web.[53]

Together with technological innovation, scientific creativity using big data oriented to the social dimension (economic, societal, cultural, political, etc.) of the cloud can be highlighted. It has been increasing over the years, especially at the macro level, and is now a particularly high source of income in big corporations, such as Amazon[54] or Facebook.[55] In this regard, 'cloud-computing giants such as Amazon Web Services, Microsoft Azure and Google Cloud offer standardised product for their corporate customers' (The Economist 2019c, p. 54). Meanwhile, practical applications (apps) have developed a polyhedral creativity (commerce, tourism, etc.) and have also extended the range of action between users of the micro (individuals and groups) and meso (companies and organizations) levels.

1.4 Creativity and Three Groups of Scientific Disciplines

Philosophically, creativity is related to terms and statements that involve some kind of difference and, ordinarily, a positive component in comparison with the previous situation. Thus, creativity is related to characteristics such as originality, imagination, novelty, inventiveness, new vision, some ability for improvement, capacity for new synthesis, openness to a manifold of opportunities, etc.[56] These features can be found in science, where creativity always involves some kind of novelty, which might be longitudinal (horizontally expandable) or transversal (vertically novel or genuinely new).

Novelty can be in four aspects of *scientific activity* as human undertaking: (a) in the aims of scientific research, (b) in the processes used by science, (c) in the results obtained, which might be different than previous ones or obtained through better means, and (d) in the consequences that follow the results achieved. In this regard, an expression such as 'innovative creativity' is *de facto* redundant,[57] insofar as creativity *eo ipso* always seeks innovative factors, which might be structural or dynamic (or even pragmatic).

From a philosophical viewpoint, creativity can be present in at least four different methodological *levels* of the scientific enterprise: (1) in the development of science, in general, insofar as it offers something that other human activities cannot achieve; (2) in the research made in groups of sciences, either formal or empirical (natural, social or the artificial), because they can add features that are different from other groups of disciplines; (3) in the elaboration of a specific science (physics, economics, pharmacology, etc.), which can solve problems that were unsolvable in other disciplines, due to the type of problem or the characteristics of the object studied; and (4) in the agents (individuals, communities, organizations, etc.) while doing science, because science is *our* science[58] and it is developed according to human categories.[59] These agents, who have a key epistemological role, can contribute top-bottom or bottom-top to the research, depending on the type of methodological development they wish to carry out.

Also, creativity depends on the kind of scientific undertaking to develop, which may be basic science, applied science and application of science. In the case of basic science, creativity is linked to scientific explanations, in their various forms (supported by laws, probabilities, causes, ends, etc.), or to scientific predictions (whether in terms of foresight, prediction, forecasting, etc.).[60] In applied science, creativity can accompany the elaboration of predictions, either in the procedural phase or in the more mature period of scientific methods, and prescriptions, to offer the patterns or appropriate guidelines once the possible future is anticipated. In the application of science, creativity is again involved and is more circumscribed. This time, creativity is delimited by the context of use, where prediction and prescription must be combined to solve the specific problems raised in a wide range of environments.

Commonly, the creativity of the scientific side of the network of networks relates to applied science and application of science. It is developed through three groups of scientific disciplines: (I) sciences directly related to the Internet (understood in the broad sense), such as Network science or Web science;[61] (II) sciences that use the layers or ontological strata to develop new aspects, as economics or communication sciences usually do; and (III) sciences that use data and information that emerge from the use of the network of networks, as science data does.[62]

Some of these disciplines are new sciences (mainly in the first and third groups), which are generated in this domain of the artificial and with social projection, while others are an extension of existing sciences (above all, in the second group). In recent decades we have seen that there is a feedback, or bidirectional, relation between applied science and application of science, which leads to a dynamic of constant creativity in the Internet (in the broad sense) to deal with the problems and the search of pragmatic solutions to specific problems of the users of the network of networks. In this regard, 'scientific achievement does, in fact, involve creativity' (Barrett et al. 2014, p. 50).

1.5 Technological Innovation and Scientific Creativity in the Network

How technological innovation interacts with scientific creativity requires first recognizing the presence of innovation in each of the technology components. Thus, (i) new language comes with technological innovations: new terms for novel realities and new content for old terms (such as radio or television, which are now digital and might be worldwide). (ii) The technological system is innovated on the basis of its operativity (e.g., in the background technology for the network and the information and communication technologies [ICT] of the users) in order to achieve new objectives (e.g., higher connectivity or better security). (iii) The technological knowledge seeks new contents to achieve success in the creative transformation of the reality.[63] (iv) The technological methods involve an active change of the previous processes while looking for new goals, which are commonly more sophisticated than the existing ones. (v) The technological objectives follow a dynamic of superseding the available results, thinking of new products and artifacts for the agents (individuals, groups, organizations, etc.). (vi) The existing values of technology, both internal (effectiveness, efficacy, etc.) and external (social, cultural, ecological, ergonomic, political, etc.), are also under scrutiny in favor of the demands on behalf of the internal technological improvements and the new level of aspirations of the public that uses such technologies.[64]

Unquestionably, innovation in technology involves some kind of novelty,[65] which is remarkable in the case of the Internet conceived in the strict sense (i.e., the network). Novelty may appear in four phases of technological work: (I) in the *design* of the possible product or artifact; (II) in the *processes* proposed that undertake to achieve such results; (III) in the actual *result* of the technological track (i.e., the product or artifact obtained); and (IV) in the *consequences* that follow from the result produced or obtained. Over the years, we have seen that in the network there has been novelty in the design, in the processes, in the results, and in the consequences.

Thus, technological innovation in the network reflects one of the central features of the transformative activity that is characteristic of the human technology and involves creativity, insofar as it seeks new ends and means to achieve the desired effects (a new product or artifact) and a range of consequences.[66] In this regard, new technologies related to the network are possible[67] (as is the case with the information and communication technologies) or improvements in the available technologies (e.g., the digital enhancements of radios and televisions). These aspects are often recognized through the mechanisms to evaluate inventions, which, in a number of cases, involve patents.[68] Furthermore, there are features of the *novelty* of technological innovation that are particularly relevant.[69]

(I) Scientific creativity commonly has a direct influence on the novelty of the technological *design*, insofar as scientific knowledge (*know that*) is needed for making designs in technology (such as ICT), in addition to the specific technological knowledge (*know how*). Furthermore, creativity can also have a role in the evaluative knowledge (*know whether*), which uses rationality to choose the preferable goal of one technology from the possible objectives considered.[70] Among the usual reasons for the technological change are social preferences and the new aspirations of society.[71] The origins of the network were different, as it was first conceived for military purposes, then for use by universities and, later, by the general public.

Concerning the designs, novelty in a technological innovation depends on the aims (e.g., something completely new or different from the existing technological items) and their values, as well as on the knowledge – scientific as well as specifically technological – used to attain them. Thus, evaluative rationality is also involved here, which is the rationality that deals with ends.[72] This is because, in technology, there are internal and external values (as well as endogenous and exogenous ethical values) that are relevant for the selection of novel goals and designs. As is well known, values may change over time, especially external values. But these initially external values, as in the case of sustainability, can later be internalized.[73]

(II) As regards the *processes*, the novelty of the technological innovation can be different from the novelty of the technological progress. As a matter of fact, not all innovation (technological, social, etc.) is *eo ipso* a step forward in the right direction, which explains the insistence on concepts of change such as 'sustainable development.'[74] *Progress* is not, in principle, a neutral concept; it always involves a positive side and looks especially to the final outcome of an activity or undertaking, whereas the processes of innovation can bring about novelties that are not positive (or not positive enough)[75] and, then, new technological alternatives can be developed.

(III) Novelty in the technological *results* can be connected with two main kinds of changes: (i) the refinements of current products or artifacts, i.e., the longitudinal line, and (ii) the innovative new products or artifacts, i.e., the transversal line. Both require a kind of heuristic – a search regarding technological problem solving – which is more positive in the second case than in the first. 'Innovative new products or artifacts' can be the outcome of novel designs or new processes according to a 'positive heuristic,', such as devices for improving the network in issues related to bandwidth, speed, stability, connectivity, etc. Meanwhile, 'refinements of current products or artifacts' have a somehow 'negative heuristic' regarding the key aspect of that specific technology, while improving other aspects of that technology (e.g., the support for digital radio or digital television).

If we consider the criteria of 'novel facts' that come from the analysis of Imre Lakatos' approach and use them to understand novelty in the realm of technology (in this case, the network), we find some interesting distinctions. These involve the idea of change in comparison to previous or posterior stages. In addition, they assume, in principle, the existence of a variation with some improvement, which is also implicit in the characterization of innovation. Although six options have been considered already,[76] they can be reduced *de facto* to four possibilities,[77] which are adapted here to the traits of technology:

(1) A *strictly temporal* novelty in technological products or artifacts is when nobody before had been able to do something similar. Thus, the

product or artifact was completely unknown when the design was proposed. (2) A *heuristic view* of technological novelty is when the product or artifact at stake was not used in the technological design. (3) A technological novelty with respect to *previous designs* might arise when the product or artifact was not predicted by the best existing predecessor to the design available, or it might be a product or artifact that has no counterpart among the consequences of the predecessors to the design. (4) A temporal novelty for the *individual*: the product or artifact is unknown to the person who constructed the design at the time the theory is constructed, or it is novelty with respect to *design* insofar as that was not something expected to be produced on the basis of such design.

(IV) As regards the *consequences*, it seems clear that the processes of technological innovation, such as the Internet in the strict sense, can be socially evaluated also (in terms of society, culture, economy, etc.), mainly because of their possible consequences. 'Sustainable development' is an interesting notion in this regard, because it is related to multiple technological processes. Furthermore, it connects with the analysis of what kind of technological possibilities should be actualized, for example, regarding the problem of the future of the Internet as a whole,[78] including the three layers mentioned here,[79] or the future of each of the main layers (the network, the Web and the cloud computing, apps and mobile Internet).[80]

Sustainable development combines internal terms (such as a cognitive content or some procedures or methods)[81] and external ones, due to the expected results and the social consequences of linking human beings with technology and their being interwoven with the environment (the natural, social or artificial settings). It is a notion that includes empirical contents (some of them related to applied sciences) and value premises (social, cultural, political, economic, etc.). Even so, 'sustainable development' raises the relevant question of the development of technological processes that can cause some harm (e.g., in terms of privacy in the case of the network or problems regarding cybersecurity of the Web or of cloud computing and apps).[82]

1.6 Creativity and the Social Innovation on the Layers of the Internet as a Complex System

Although designers have a role regarding social innovation, it comes primarily from the constant relationship of users (individuals, groups, organizations, etc.) with the network of networks, which seeks new aims, processes and results in an environment embedded in historicity (in science, technology and society).[83] The social dimension of the

Internet – in the broad sense – has been reflected in the changes in the technological infrastructure, in the developments in the Web[84] and in the novelties brought about by cloud computing, practical applications (apps) and mobile Internet.

Therefore, in addition to the social dimension of the Web and cloud computing, along with apps and mobile Internet, there is the social dimension corresponding to the network, which connects with creativity regarding technological infrastructure (mainly background, service provider and ICT), because the creativity of human agents (individuals, groups, organizations, etc.) has *versatility* regarding objectives, undertakings and products or artifacts (with their ensuing consequences) of technology, in general, and of information and communication technologies (ICT), in particular. This versatility is something that these technologies elude when they merely carry out the tasks for which they have been originally designed.

Again, the source of creativity can be 'internal' and 'external.' When technological innovation is raised from an exclusively internal point of view (commonly with the support of Artificial Intelligence, in the case of ICT), then it is configured as accumulated learning from the previously available information. This performance can only lead to derivative products, instead of giving rise to authentic creativity, capable of generating a genuine novelty (i.e., something novel transversal or with vertical novelty). Therefore, technological innovation should be open to the external side. This includes human agents (technologists, scientists, users) who, driven by the social dimension, expressly seek creativity in technological design (e.g., for cybersecurity, connectivity, accessibility, etc.) and can end up creating a startup.

What has happened to the creativity of the social dimension in the last decade was already predicted in 2008: 'technology providers will adapt their products to the manner in which they are taken up by the users. Furthermore, technology providers will learn from users. Moreover, the way users employ new technologies may change the way innovations were conceived by the technology providers. All such kinds of 'user-involvement' have been seen in the case of the Internet, which illustrates that a technology-determinist approach will not be able to understand the development of the Internet and how it has come to be what it is today' (Henten and Tadayoni 2008, p. 47).

An important part of the social dimension now looks to the Internet of Things (IoT). First, after the Second World War, large corporations and governments developed Information and Communication Technologies (ICT). Then, these ICT progressively reached the general public through desktop PCs, laptops and smartphones. In this new phase, it 'will bring the benefits – and drawbacks – of computarisation to everything else, as it becomes embedded in all sorts of items that are not themselves computers, from factories and toothbrushes to pacemakers and beehives' (The Economist 2019d, p. 3). On the one hand, there is the cost and technological difficulty that accompanies the Internet of Things;[85] and, on the other, there are security problems for users when they have a world of ubiquitous sensors, because experience teaches that 'thirty years of hacks and cyber-attacks have proved that computers are insecure machines' (The Economist 2019d, p. 4).

1.7 Creativity and Communication through the Network of Networks: The Problem of Fake News

Today, one of the most important problems of the network of networks is fake news. This problem requires a combination of the three factors analyzed here: scientific creativity, technological innovation and social dimension. (1) Scientific creativity is required to deal with the new contents, which mainly affect the second – the Web – and third layers (especially mobile Internet). The challenge is to achieve communication in which the search for objectivity prevails, to be able to reach true content or, at least, to discard clearly false or, biased content. (2) Technological innovation is needed regarding the two main sides of the problem: in the machines available and in the use made by the people. First, we need to detect and stop bots and new artifacts that use Artificial Intelligence for their operations of spreading false or biased information.[86] This can run into the problems of states that allow or promote the malevolent use of these devices in their territory. Second, adequate mechanisms ought to be in place so that, while guaranteeing the neutrality of the network and the necessary freedom of expression, if users are producers and disseminators of content that is patently false (misinformation) or discernibly tendentious (disinformation), the social damage can be avoided or reduced. This can be achieved by limiting a repetition of sending (as WhatsApp does now).[87] (3) The social dimension of the problem is evident, as we have seen with the anti-vaccine movements or in the dissemination of news about Covid-19.[88] Some social networks, when the problem reaches a special magnitude, take exceptional measures.[89]

From a methodological point of view, the problem of fake news communication poses three main moments,[90] each of which requires creativity: (i) prevention to preclude the emergence of fake news, (ii) detection to reveal its presence once it has been put into circulation and (iii) the task of correcting or eliminating fake news. For this type of methodological task, it is necessary to know how to combine two major components: (a) the sciences related to the network of networks, mainly Web science, and the communication sciences, as applied sciences of design; and (b) the social dimension of the Web, so that it is at the service of society instead of social division or disintegration. Within the communication sciences, it is necessary to know how to combine their characteristic of applied sciences and their role for the application of science.

To claim that the problem of fake news can only be solved with the use of data science, which is primarily quantitative, and Artificial Intelligence – to the extent that it is still primarily syntactic in its *modus operandi* – seems certainly naïve. Indeed, the diversity of forms of fake news (mainly misinformation and disinformation) and the complexity of the network of networks – which is structural, dynamic and pragmatic – requires more creativity than that derived from quantitative approaches and syntactic priority. It is necessary to use a set of scientific disciplines with well-defined aims and coordination among them, together with the corresponding technological support and constant attention to the historicity of the social dimension.

First, sciences are required to be able to anticipate the sources that can originate fake news, in terms of both misinformation and disinformation. (I) Misinformation is erroneous information; it is a statement that expresses a false proposition, an image or a sound that does not reflect something real but a fictitious content. Within the empirical realm, this concerns occurrences in nature, social events or artificial phenomena. (II) Disinformation is a statement that conveys a deliberately erroneous proposition. It is, strictly speaking, a lie (where the intention to deceive is an essential component), which is usually disseminated to mobilize (individuals, groups, organizations, institutions, etc.) in a certain direction. Disinformation can also be audiovisual, which involves some degree of image or sound manipulation.[91] These two forms of fake news are related not only to human agents but also to the use of Artificial Intelligence, in particular bots.[92]

Second, scientific knowledge should help to detect the presence of fake news. There, internal aspects and external elements of human knowledge come into play, which are usually combined in mechanisms such as fact-checking. Qualitative factors are also at stake, in addition to quantitative ones. To assess what is true and what is false – and the semantic, epistemological and ontological elements leading to that assessment – goes beyond the field of a single discipline (be it computer science or any other), since an interdisciplinary view is needed, where theoretical schemes allow us to grasp the complexity – structural, dynamic and pragmatic – of the information that transits through the network of networks, especially the Web and mobile Internet. In this field of delimitation of what is true and what is false, the contribution of philosophy is key.[93]

Third, the elimination of fake news, so that it does not reach the public and does not cause harmful effects on society, is the decisive social challenge in this regard.[94] To achieve this result, the differences between locutionary, illocutionary and perlocutionary acts must be clearly understood, since their impact on the network of networks – especially their social dimension – varies considerably. The mere communication of a statement is not the same as transmitting a statement that has associated social or formal effects or that the statement is used for effective practical action, through the tasks of dissuading, persuading, etc. This involves three successive levels: (a) making false statements or audiovisual content disappear; (b) managing to eliminate statements or audiovisual content that have formal effects in social life (such as swearing or promising a misled position); and (c) managing to discard biased messages that include manipulation of social actions (for example, in an electoral process).

1.8 Origin and Characteristics of This Book

Looking at this book as a whole, it can be placed in the coordinates or central elements proposed in this chapter: creativity on the Internet as a complex system, involving scientific creativity, technological innovation and social innovation. Thus, it emphasizes that there is an interaction between the scientific side, the technological facet and the social dimension in the network of networks. This leads to structural complexity (epistemological and ontological),[95] dynamic complexity (in terms of historicity)[96] and pragmatic complexity (due to the need to meet the specific needs of the system's users).

Altogether, the scientific side has a special weight in this first chapter and in the entire volume, because most of the papers have their roots in the Conference on the Internet and Science: An Analysis from Structural and Dynamic Complexity (*Jornadas sobre Internet y Ciencia: Análisis desde la complejidad estructural y dinámica*), held on March 14 and 15, 2019 at the University of A Coruña, Campus of Ferrol. This *XXIV Conference of Contemporary Philosophy and Methodology of Science* was organized by the Center for Research in Philosophy of Science and Technology (*Centro de Investigación de Filosofía de la Ciencia y la Tecnología*, CIFCYT) with the endorsement of the Society for Logic, Methodology and Philosophy of Science in Spain.

After the initial contextualization presented in this chapter, four parts organize the contents of this book: (I) Configuration of the Internet and Its Future, (II) Structural and Dynamic Complexity in the Design of the Internet, (III) Internal and External Contributions of the Internet and (IV) The Internet and the Sciences. This organization of the volume follows a path that takes relevant aspects: the present and future of the Internet as a whole, structure and dynamics traits, internal and external factors and the use that science makes of the network of networks.

Besides the epistemological and methodological aspects, this book also pays particular attention to the artificial ontology of the Internet based on creative designs. Thus, Chapter 2 is focused on 'The Internet as a Complex

System Articulated in Layers: Present Status and Possible Future'. Wenceslao J. Gonzalez (University of A Coruña) assumes the general setting of complexity of the network of networks as diversified in structural, dynamic and pragmatic. In this regard, the analysis of the Internet – in the broad sense – as a complex system considers each of the main layers mentioned here.

(1) The main components and the factors for the future of the layer of the Internet *sensu stricto* are studied. The technological facet is dominant here. The Internet as a scale-free network and its possible future are considered. After that, the Internet of the Things is seen in the context of complexity. (2) There is an analysis of the layer of the Web in several directions: (i) the future of the Web as an undertaking where scientific side, technological facet and social dimension are intertwined; (ii) the future of the Web as modulated by an interaction of internal and external perspectives; and (iii) three possible scenarios for the future of the Web based on the present trends (dystopian, open and in-between). (3) A study is made in terms of complexity of the layer of the cloud computing, practical applications (apps) and the 'mobile Internet,' based on their present status to predict their possible future. Thereafter, a coda adds some philosophico-methodological reflections on the present status of the Internet as a complex system and its near future.

Of special importance is 'The Future of the Web,' where James Hendler (Rensselaer Polytechnic Institute, Troy, New York) – a key figure of the Web science, in general, and semantic web, in particular – analyzes several scenarios. He is well aware of one of the biggest societal changes in human history, which began when the foundations for the World Wide Web were delineated by its inventor, Tim Berners-Lee.

Hendler emphasizes its exponential growth, which affects the world today, as we have seen during 'life online' due to Covid-19 mobility restrictions. Thereafter, Chapter 3 considers the future of the Web with three main options: (i) dystopian outcome, (ii) prospect open to beneficial inputs and values, and (iii) in-between panorama, which has characteristics of the previous possibilities. Again, there is an interconnection of scientific side, technological facet and social dimension. In this regard, Hendler recognizes that Web science was created as an interdisciplinary field to study the World Wide Web.

Structural and dynamic complexity are in the focus of the analysis of the design of the Internet. Thus, in Chapter 4 – 'Designing an Internet of Machines and Humans for the Future' – Yan Luo (University of Massachusetts Lowell) studies features of the architecture and applications of the network of networks. This complex system has undergone a transition from a small network of connected computers in laboratories to a massive number of connected things, including sensors and actuators geographically distributed in the physical world, besides the emergence of the Internet of Things.

Luo pays attention to the growing network programmability and the surge of machines related to Artificial Intelligence: (I) the network has evolved from an initial design to a highly customizable and programmable one, which makes operation and management increasingly more efficient and cost-effective; and (II) there is a significant trend of big data and data driven machine learning, which are the bases for applications for disease diagnosis, medicine discovery, computer vision, autonomous driving, etc. This chapter provides an overview of the structure and dynamics of these phenomena and analyses the underlying enabling factors. The central aims of the paper are to outline the general trend in the Internet design conception and to provide some predictions for the new Internet that serves the future human society.

Another angle of analysis, directly focused on dynamic complexity, is available in Chapter 5: 'Strategies for Managing Dynamic Complexity in Building the Internet.' Ole Hanseth (University of Oslo) proposes a design theory for information infrastructures (shared, open, heterogenous and evolving systems of information technology capabilities). This includes the Internet and Electronic Data Interchange (EDI) networks. The dynamics of information infrastructures is non-linear, path-dependent and influenced by network effects and unbounded user and designer learning.

Hanseth deals with two problems related to designs of information infrastructures: (a) the bootstrap problem (to directly meet the user's needs in order to be initiated) and (b) the adaptability problem (local designers facing the information infrastructure's unbounded scale and functional uncertainty). Both problems are discussed based on complex adaptive systems theory to derive information infrastructures' design rules (i) to address the bootstrap problem by generating early growth through simplicity and usefulness and (ii) to solve the adaptability problem by promoting modular and generative designs. He illustrates these principles based on the interpretation of the history of Internet: the emergence and early growth of the Internet, for the first problem, and the transition of the computer networks connecting universities to the Internet in the Nordic countries, for the second problem.

Internal contributions based on the scientific side of the network of networks are developed in 'Data Observatories: Decentralised data and interdisciplinary research.' In Chapter 6, Thanassis Tiropanis (University of Southampton) looks at observations and data in an artificial sphere, distinct from the natural and social environment. In this regard, methodologies of data analysis and digital platforms can provide observations on those data. The use of methodologies of interdisciplinary areas, such as Web science and Internet science, afford a meaningful analysis of data on activity taking place on the Web and the Internet.

In the Chapter 6, Thanassis Tiropanis discusses the concept of 'data observatories' as decentralized platforms for sharing observations and data. He uses the lessons learned from the design and deployment of such elements in academia over the past decade. Further, the chapter outlines an architecture for the deployment of data observatories and reviews their potential to support scientific discourse within and across disciplines. This paper argues that decentralization will have an impact outside academia as well. The chapter examines the necessary conditions of decentralization, digital literacy requirements and policies to support wider engagement in scientific discourse.

A combination of internal and external contributions is available in Chapter 7: 'Digitization, Internet and the Economics of Creative Industries.' Ruth Towse (University of Bournemouth) takes into account the need for a new economic approach to creative economy in the digital era. Economics is one of the sciences that has undergone the greatest development, both in its artificial aspect – new designs – and in its social dimension, thanks to the use of the network of networks. Thus, through technological support, there are digital developments that have a clear social impact and lead to creative industries. This is the case of everything related to copyright, which affects a wide range of products (books, journals, music, television, cinema, etc.), and where economics intersects with law.

Towse sees in the new economic approach a reorientation towards the production and consumption of intangible, ephemeral goods produced and distributed electronically. The old economic theories were mainly focused on the application of welfare economics to the performing arts and heritage sectors, whereas the new economic models (such as subscriptions to large catalogues of music, films and broadcasts, enabled by the Internet) have a huge impact on the cultural environment. Towse considers the labor market of creators and performers as well as the impact of Covid-19 in the arts and creative industries, taking into account the contribution that the Internet can make to future consumption and production of the arts.

External contributions are the focus of Chapter 8: 'Digital surplus: Work in the Information Age'. Enrique Alonso (Autonomous University of Madrid) looks at the technological facet and the social dimension of the network of networks. He considers it as the contemporary factor that modulates many human activities, assuming that labor is affected in many aspects, such as a paid profession, a type of human task that can now be carried out, benefits derived from the new support in terms of contents, social relations created, mobility accessibility (including teleworking), etc. All of this means that digital work represents an added value in the knowledge society. From a social point of view, this new situation generates different attitudes, some more realistic than others.

Alonso sees an increase in complexity regarding the concepts of applied economics in this digital environment. He analyzes the mechanisms to recover data that the users generate on the digital platform and wants to open the door to free access to non-public data through duly standardized procedures, which should be monitored by public institutions. In this regard, he indicates the possibility of demanding an economic return on the benefits obtained by the platforms on which the users of the network of networks generate content. In addition, the paper anticipates the emergence of new economic activities that are in line with the possibilities of the Information Age.

Twofold contributions of the Internet to the sciences are developed in the next chapters. In 'Biology and the Internet: Fake News and Covid-19,' Wenceslao J. Gonzalez (University of A Coruña) analyzes, from a philosophico-methodological point of view, two interconnected aspects. First, he reflects on the issue of the status of biology in the context of the relationship between the Internet – understood in the broad sense – and science. This study involves two successive steps. I) Scientific development and dissemination of the knowledge are obtained as the two main forms of relationship between the sciences of biology and the network of networks. This complex system of the Internet is composed of the technological infrastructure, the Web and the cloud computing along with practical applications and mobile Internet. II) The relationship with biology in the Web layer, which is both an instrumental and a substantive contribution to the sciences of biology (biochemistry, molecular biology, genetics, microbiology, botany, zoology, ecology, etc.).

Second, this Chapter 9 addresses the issue of fake news surrounding Covid-19 and sees it within the context of the relationship between biology and the Internet. This sphere includes its connection to issues of vaccines and the safety of their use. This task is done in three steps: (1) the list of the main problems related to this issue of fake news and Covid-19 that have been identified to date. (2) Consideration of misinformation and disinformation as the two most relevant forms of fake news that require attention in this case. (3) The philosophico-methodological routes to be followed to deal with the problems of fake news related to Covid-19. Afterwards, the paper includes a coda that points out a set of relevant aspects in the relationship between scientists as advisers and the development of policy making.

Thereafter, Maria Jose Arrojo (University of A Coruña), in her chapter 'The Novelty of Communicative Design on the Internet: Analysis of the Snapchat case from the Sciences of the Artificial,' exemplifies how the communication sciences are design sciences, in addition to being social sciences. The paper sees this duality – artificial and social – in the use of the network of networks for communication purposes, again in the Web layer, where new communicative designs are made for different styles of communication. This is clearly the case in the social networks. In this regard, from the point of view of the sciences of the artificial (with aims, processes and results), Chapter 10 considers the novelty of the social network of Snapchat.

Methodologically, the novelty of the communicative design is due to two factors: (a) the scientific creativity to develop a design and (b) the technological support of the design to be operational and achieve the proposed goals. Another feature is the constant feedback between the scientific creativity of communicative designs and the technological innovation developed through the network of networks. Furthermore, the communicative objectives, together with the artificial communicative design, requires the social dimension of communication, in general, and of the Internet, in particular. There are also the epistemological and ontological factors. Thus, complexity is observed in several ways: (i) in the constant interaction between scientific creativity and technological innovation; (ii) in the combination of artificial and social elements in communicative designs; and (iii) in the permanent intertwining of the structural level of communicative designs and the external factors conditioned by the environment in which they are developed.

About the conference mentioned in this section, allow me to express my recognition to the persons and institutions that cooperated in the original event organized by CIFCYT. First, my appreciation to the speakers at the *Jornadas* who have contributed to this volume: James Hendler, Yan Luo, Ole Hanseth, Thanassis Tiropanis, Ruth Towse, Enrique Alonso and Maria Jose Arrojo. I am really pleased with the efforts they made regarding their papers then and with the final versions for this book. Second, my acknowledgement to the organizations that supported this conference: the University of A Coruña (especially to the Rector of the University and the Vice-rector for Economics and Strategic Planning), the Santander Bank, the Foundation Philosophy of Science and Technology, and the Society of Logic, Methodology, and Philosophy of Science in Spain. Third, I am grateful to the technical secretary, Jessica Rey, and the support team of the conference.

Regarding this book, in addition to the collaborators in the volume, let me point out that I am thankful to Amanda Guillan, Alba Garcia Bouza and Ana María Alonso for their contribution to this book. Finally, my gratitude to the Centre for Philosophy of Natural and Social Sciences (CPNSS) at the London School of Economics, where I have carried out a substantial part of the research that I publish here.

Notes

1 This paper has been written within the framework of research project PID2020-119170RB-I00 supported by the Spanish Ministry of Science and Innovation (AEI). An initial version of this text was presented at the Congress on *Creativity 2019*, 1st World Congress of the Brazilian Academy of Philosophy, held at the Federal University of Rio de Janeiro on 8–13.12.2019.
2 See, for example, Corrales et al. (2017); Park (2019); Lynch (2016); and Gonzalez (2020c).
3 Alongside technological and social innovation processes are the organizations that make innovation based on creativity viable. See, in this regard, Amar and Juneja (2008). On the main approaches to characterize 'innovation' (agents, results and processes), see Echeverria (2017), p. 80.

4 'Creative work can be found in any job that requires complex, ill-defined problems where successful performance depends on the generation of novel, useful solutions' (Barrett et al. (2014), p. 39).

5 Cf. Gonzalez (2023a).

6 On creativity related to the elaboration of content in the context of cultural production, see Duffy et al. (2019). Content creation is a rather generic task, in that it is not specifically linked to the solution of concrete problems, whereas scientific creativity is directly linked to the resolution of current or possible well-defined problems. This difference is similar to that between 'design' and 'scientific design.'

7 Cf. Gonzalez (2018a). See also Gonzalez (2023b).

8 Cf. Clark (2018), p. 37. It should be emphasized that a good design must incorporate internal and external values. The former, in a complex system, aims at effectiveness and efficiency. The latter must take into account the social dimension. When a design is well done, 'designers teach a fresh way of thinking that calls for heightened sensitivity to human needs, great empathy for the people who use technology, and increased willingness to engage with stakeholders as partners and participants' (Shneiderman 2019, p. 1383).

9 Cf. Schultze and Whitt (2016); and Gonzalez (2019).

10 Cf. The Economist (2018).

11 An analysis of these three main layers is available in Gonzalez (2023a).

12 On the features and uses of 'creativity', see Boden (1994); Csikszentmihalyi (1996); and Kaufman and Sternberg (2010).

13 This epistemological and methodological pluralism is, in principle, a feature that accompanies the philosophical study of the infosphere. On the characterization of the infosphere, see Floridi (2014), pp. 40–41. See also his general approach to information: Floridi (2011).

14 Other aspects of technology innovation – semantic, logical, axiological and ethical – are included in Section 4 of this paper.

15 Cf. Yap et al. (2020).

16 'The Internet is more than TCP/IP (Transmission Control Protocol/Internet Protocol). The basic definition of the Internet is communication using TCP/IP. However, what has come to be understood as the Internet encompasses a host of different technologies, e.g., the World Wide Web (the Web), compression technologies, streaming technologies, etc., where the different technological elements have different implications for the media' (Henten and Tadayoni 2008, pp. 46–47). Since that year, a new layer, or stratum, has been developed based on the Internet as technological platform, where cloud computing and practical applications (apps) are located.

17 Cf. Gonzalez (2022).

18 On these three components and, in particular, the scientific approach, see Gonzalez (2018a).

19 This includes the development and use of Artificial Intelligence. On AI in the context of design sciences, see Gonzalez (2017).

20 Cf. Gonzalez (2013a).

21 This characteristic of active participation is particularly clear in the cases of the Web and the apps.

22 Cf. Gonzalez (2013a), pp. 15–16.

23 In addition to technological innovation, there is social innovation. Along with innovation by experts, there is also innovation by users. Cf. von Hippel (2005).

24 On these components of technology, see Gonzalez (2005), pp. 3–49; especially p. 12.

25 On the history of the Internet, there are numerous syntheses, among them Leiner et al. (2009); Clark (2018), pp. 16–29; Cohen-Almagor (2011); and Elton and Carey (2013). From the point of view of the Web, see Berners-Lee (2000) [1999].

26 Before the Internet as a technology platform, there were several forms of electronic commerce. Cf. Gonzalez (2020a). This previous stage in the case of electronic commerce was 'in sectors such as retail automotive, electronic data exchange (EDI) for application-to-application interaction is being used regularly. For defense and heavy manufacturing, electronic commerce lifecycle management concepts have been developed that aim to integrate information across larger parts of the value chain, from design to maintenance, such as CALS (Computer Assisted Lifecycle Support or Computer Aided Logistics Support)' (Timmers 1999, p. 3).

27 About bitcoin as an alternative to the current monetary and payment system, cf. Weber (2016). 'Bitcoin was originally sold on the promise of unending the global monetary system. Its success now hinges on finding a more modest role within it' (The Economist 2021b, p. 9).

28 A number of these limits can be seen within the framework of the general limits of technology connected to the limits of science, cf. Gonzalez (2016). From the point of view of users, the problem of limits has another side, cf. Arbesman (2016).

29 'A network that is maximally general with respect to the fundamental impairments (a theory of stability) and can evolve in response to impairments that change over time (a theory of innovation) will be long-lived. (…) the Internet is a good example of a general network by this definition and see its longevity as a consequence of that fact. The use of packets as multiplexing mechanism has proved to be very general and flexible mechanism. Packets support a wide range of applications and allow for the introduction of new technology as it evolves' (Clark 2018, p. 173).

30 This historicity is related to the dynamic complexity of the Internet and its future development as a technological platform based on the support of design sciences. Cf. Gonzalez (2018b).

31 Cf. Clark (2018), p. 179.

32 'Problem exploration—recognizing, framing, and defining a need—has been identified as a critical component of design process' (Murray et al. 2019, p. 249).

33 Such external pressure for creativity in designs can come from problems posed by management of organizations, clients of business firms or product users. Cf. Murray et al. (2019), p. 249.

34 In some cases, they operate *de facto* as a monopoly in the field of services they offer, frequently buying smaller companies that may be competitors. On this issue, see The Economist (2019e).

35 Cf. Greenstein (2015). In this regard, see The Economist (2021a), p. 10.

36 The change from within comes from a change of focus. This implies a modification in the culture of the company or organization in order to meet the new social demand or create it. In this regard, Microsoft

> shift away from a focus on the Windows operating system and towards Azure, Microsoft's cloud-based services offering. It involved a willingness to let programs run on Apple and Android smartphones, something the company had previously avoided. However, the turnaround also required a change to the company's culture and that is the main subject of the LBS [London Business School] study. The shift to cloud-based services meant that revenues would be generated in a different way. (…)

> The sift was enormous. Around 40,000 people had to change how they did their jobs.
>
> (The Economist (2019a), p. 51)

37 Cf. Floridi (2014), p. 18.
38 Cf. Gonzalez (2020b).
39 On these three possibilities of scientific activity, see Gonzalez (2015b), pp. 32–40.
40 About the other components of technology, see Gonzalez (2013a), pp. 21–22.
41 When Paul Humphreys analyzes emergence from a general point of view, his focus is on novelty: 'Emergent features result from something else, they possess a certain kind of novelty with respect to the features from which they develop, they are autonomous from the features from which they develop, and they exhibit a form of holism' (Humphreys 2016, p. 26). Thus, he analyzes these four features: (1) relational, (2) novelty, (3) autonomy and (4) holism. But, for him, 'holism is insufficient for emergence because holistic properties can be possessed by systems that do not display emergent features' (Humphreys 2016, pp. 36–37).
    His analysis of novelty includes five kinds of conditions, cf. Humphreys (2016), pp. 29–32. They are (a) novelty of an entity with respect a theoretical base; (b) novelty regarding a given explanatory base; (c) novelty with respect to a base set of entities; (d) causal novelty regarding a domain; and (e) novelty when 'different laws apply to emergent features than to the features than to the features from which they emerged' (Humphreys 2016, p. 32).
42 Cf. Dasgupta (1994).
43 Cf. Dasgupta (1996).
44 Cf. Clark (2018), p. 145.
45 The human experience regarding the network is based upon the common technological standards (TCP/IP) used in the Internet as packet transport, and they require the supporting technologies (broadband, ethernet, Wi-Fi, etc.). The users need an adequate hardware for their ICT in order to get the connection to the Internet Infrastructure Provider. Cf. Clark (2018), p. 7.
46 Cf. Clark (2018), pp. 5–6.
47 Cf. Ramadani and Gerguri (2011).
48 In this regard, Jobs took an idea from Alan Kay (Xerox Corporation): 'The best way to predict the future is to invent it' (Isaacson 2011, p. 95). Alan Kay worked at the Palo Alto Research Center or Xerox Park.
49 Cf. Gonzalez (2013a), p. 19.
50 The NeXT computer used in 1989 by Tim Berners-Lee in the initial design of the Web has been exhibited at the Science Museum in London during 2019, within the framework of the Information Age.
51 Cf. Hendler (2023), pp. 71–73.
52 'The following stage of the World Wide Web was dubbed Web 2.0, a term first coined by [Darcy] DiNucci (1999) and years later popularized by [Tim] O'Reilly (2005), which refers to a new dominant form of interaction between people through different Internet-connected devices' (Guilló 2015, pp. 42–43).
53 Apps 'rely on the same web architectures' as web browsing. See Hendler and Hall (2016), p. 704.
54 Cf. Novet (2019). See also The Economist (2019b), pp. 53–54.
55 In the case of Azure, Microsoft's cloud-based services offering, the success has resulted from a shift: 'cloud services are paid for on a metered basis; revenue comes in only when customers use them' (The Economist 2019a, p. 51).
56 Cf. Gonzalez (2013a), p. 14.

57 'But his more important goal, he [Steve Jobs] said, was to do what [Bill] Hewlett and his friend David Packard had done, which was create a company that was so imbued with innovative creativity that it would outlive them' (Isaacson 2011, p. xix).

58 On this human dimension of a cognitive kind, see Rescher (1992).

59 Cf. Gonzalez (2013a), pp. 14–15.

60 On the distinction between foresight, prediction, and forecasting, see Gonzalez (2015b), pp. 68–72.

61 Cf. Tiropanis et al. (2015); and Berners-Lee et al. (2006).

62 Cf. Cao (2017).

63 Cf. Fleming and Sorenson (2004), p. 926.

64 Cf. Gonzalez (2013a), p. 22.

65 This section develops a set of ideas based on Gonzalez (2013a), pp. 19–21.

66 Cf. Gonzalez (1997); and Gonzalez (2015a).

67 Actually, there have been a number of new technologies since Arpanet and David Clark considers that what we have is *an* Internet. According to Clark, 'it was very important for the success of the Internet architecture that it be able to incorporate and utilize a wide variety of network technologies, including military and commercial facilities' (Clark (2018), p. 62).

68 Patents are a characteristic recognition of technological innovations. Cf. Ordoñez (1992). See also Fleming and Sorenson (2004), p. 909.

69 To some extent, there are traits shared by the concepts of 'innovation' and 'development.' Thus, in addition to positive aspects, there may be negative ones. This twofold possibility has led to notions such as 'sustainable development' (and an equivalent could be 'assumable innovations'). In recent years, some authors have paid more attention to negative traits. Cf. Wachter-Boettcher (2017); Eubanks (2018); and Broussard (2018).

70 In the technological designs, three kinds of knowledge can intervene: scientific, specific of technological and evaluative, cf. Gonzalez (2015a), pp. 4–11.

71 On the social dimension of technology, see Elster (1983). Regarding the social dimension of the Internet, see Winter and Ono (2015). Along with this external dynamic of technological change, there is usually an internal dynamic of the technology itself to bring about change.

72 On evaluative rationality, cf. Rescher (1999), pp. 79, 81, 92, and 172. See also Rescher (2003).

73 Cf. Gonzalez (2015a), pp. 6–20.

74 Cf. Niiniluoto (1994).

75 'Science alters inventors' search processes, by leading them more directly to useful combinations, eliminating fruitless paths of research, and motivating them to continue even in the face of negative feedback' (Fleming and Sorenson 2004, p. 909).

76 See, in this regard, Hands (1991), pp. 96–99; and Backhouse (1997), p. 115.

77 Cf. Gonzalez (2014), pp. 1–25; especially, pp. 14–16. See also Gonzalez (2015b), pp. 113–114.

78 On the future of the Internet, see Clark (2018), pp. 301–327, and the papers in the volume edited by J. Winter and R. Ono, *The Future Internet: Alternative Visions*, published by Springer in 2015.

79 Cf. Gonzalez (2023a).

80 After several decades of intense creativity in the second layer (the Web), there now seems to be a particular creativity in the third layer, such as in mobile cloud computing. It enables a wide variety of devices (PCs, laptops, tablets, smartphones, etc.) to access types of utility programs, to store information and to develop applications over the Internet, via services that are offered by cloud

computing providers. Examples of these utility programs (or Software as a Services, SaaS) include Google Apps, Microsoft Office 365 and OnLive. Cf. Liu et al. (2015). See also Mohapatra et al. (2015); and Assuncao et al. (2015).

81 On the distinction between procedures and methods, see Gonzalez (2015b), pp. 250–252.

82 Technological innovations play a key role regarding these issues. See, for example, Varghese and Buyya (2018); Vasuki et al. (2018); and El-Sofany and El-Seoud (2019).

83 An analysis of the historicity in the social dimension of the network of networks is found in Gonzalez (2020b).

84 See, for example, Berners-Lee et al. (2006), p. 770; and Hall et al. (2016).

85 'To create an IoT you need more than just a trillion cheap computers. You also need ways to connect them to each other' (The Economist 2019d, p. 4).

86 'AI is radically extending the capabilities for propagation of misinformation, of biases in information distribution and decision making, and it further extends social polarization, harassment and hate speech' (Berendt et al. 2020, p. 7). An example is the Generative Pre-Trained Transformer 3, which is a language model developed by OpenAI (an Artificial Intelligence Laboratory in San Francisco):
'GPT-3 was trained on an unprecedent mass of text to teach it the probability that a given word will follow preceding words. (...) Access to GPT-3 is restricted. For one thing, says Jack Clark, former head of policy at the organization, it might be used to mass produce fake news or flood social media with 'trolling and griefing' messages' (The Economist. 2021e, p. 67).

87 This issue also requires consideration of an ethical framework and legal liability when free actions lead to physical or moral harm to individuals, organizations or the common good of society.

88 Cf. Gonzalez (2020c). From a social point of view, among the elements that can be part of the solution are the epistemic authority of science and the relevance of social values that look to the common good.

89 'On May 5th Facebook's independent content-review body, the Oversight Board (OB), issued its most anticipated ruling since it began hearing cases last year, upholding the company's decision to suspend Mr Trump's accounts' (The Economist 2021d, p. 36). Oversight Board's decision 'points to the difficult three-way balance online platforms must strike between free speech, online misinformation and real-world harm' (The Economist 2021d, p. 36).

90 In addition to the philosophico-methodological aspects of the problem of fake news related to the network of networks, which are the focus of attention here, there are external factors involved in this issue, which are social, cultural, political or legal. Among them is the elaboration of laws against misinformation and disinformation. This can give rise to legislation that, strictly speaking, does not seek to protect the population from fake news but to modulate, condition or restrict freedom of information. In this second direction, 'between March and October of last year [2020] 17 countries passed new laws against "online misinformation" or "fake information", according to the International Press Institute' (The Economist 2021c, p. 49).

91 There is an increasing concern with some uses of deepfakes and Generative Adversarial Networks, which includes legislation in 2019 in the State of California.

92 A challenge is to 'ensure that AI brings no harm to Web users and to make AI Web bots benevolent by design' (Berendt et al. 2020, p. 14).

93 Cf. Gonzalez (2021).

94 The parallel challenge is to ensure that the level of education in society is such that citizens have the capacity to actively contribute to detecting and eliminating fake news.
95 Cf. Gonzalez and Arrojo (2019).
96 Cf. Gonzalez (2013b); and Gonzalez (2018b).

References

Amar, A. D. and Juneja, J. A. 2008. 'A Descriptive Model of Innovation and Creativity in Organizations: A Synthesis of Research and Practice', in *Knowledge Management Research and Practice*, vol. 6/4, pp. 298–311.

Arbesman, S. 2016. *Overcomplicated: Technology at the Limits of Comprehension*, New York: Penguin/Random House.

Assuncao, M. D., Calheiros, R. N., Bianchi, S. and Netto, M. A. S. 2015. 'Big Data Computing and Clouds: Trends and Future Directions', in *Journal of Parallel and Distributed Computation*, vol. 79–80, pp. 3–15. DOI: 10.1016/j.jpdc.2014.08.003.

Backhouse, R. E. 1997. *Truth and Progress in Economic Knowledge*, Cheltenham: Edward Elgar.

Barrett, J. D., Vessey, W. B., Griffith, J. A., Mracek, D. and Mumford, M. D. 2014. 'Predicting Scientific Creativity: The Role of Adversity, Collaboration, and Work Strategies', in *Creativity Research Journal*, vol. 26/1, p. 39. DOI: 10.1080/10400419.2014.873660.

Berendt, B., Gandon, F., Halford, S., Hall, W., Hendler, J., Kinder-Kurlanda, K. E., Ntoutsi, E. and Staab, S. 2020. 'Web Futures: Inclusive, Intelligent, Sustainable. The 2020 Manifesto for Web Science', *Manifesto from Dagstuhl Perspectives Workshop 18262*, pp. 1–42. Available at: https://drops.dagstuhl.de/opus/volltexte/2021/13744/ Accessed at 4.5.2021.

Berners-Lee, T. 2000 [1999]. *Weaving the Web. The Original Design and Ultimate Destiny of the World Wide Web by Its Inventor*, New York: Harper.

Berners-Lee, T., Hall, W., Hendler, J., Shadbot, N. and Weitzner, D. J. 2006. 'Creating a Science of the Web', in *Science*, vol. 313/5788, pp. 769–771.

Boden, M. A. (ed.). 1994. *Dimensions of Creativity*, Cambridge, MA: The MIT Press.

Broussard, M. 2018. *Artificial Unintelligence: How Computers Misunderstand the World*, Cambridge, MA: The MIT Press.

Cao, L. 2017. 'Data Science: Challenges and Directions', in *Communications of ACM*, vol. 60/8, pp. 59–68.

Clark, D. D. 2018. *Designing an Internet*, Cambridge, MA: The MIT Press.

Cohen-Almagor, R. 2011. 'Internet History', in *International Journal of Technoethics*, vol. 2/2, pp. 45–64.

Corrales, M., Fenwick, M. and Forgó, N. (eds.). 2017. *New Technology, Big Data, and Law*, Singapore: Springer.

Csikszentmihalyi, M. 1996. *Creativity. Flow and the Psychology of Discovery and Invention*, New York: Harper/Collins.

Dasgupta, S. 1994. *Creativity in Invention and Design. Computational and Cognitive Explorations of Technological Originality*, Cambridge: Cambridge University Press.

Dasgupta, S. 1996. *Technology and Creativity*, New York: Oxford University Press.

DiNucci, D. 1999. 'Fragmented Future' *Print*, vol. 4, pp. 2–32.

Duffy, B. E., Poell, Th. and Nieborg, D. B. 2019. 'Platform Practices in the Cultural Industries: Creativity, Labor, and Citizenship', in *Social Media + Society*, vol. 5/4, pp. 1–8. DOI: 10.1177/2056305119879672.

El-Sofany, H. and El-Seoud, S. 2019. 'A Novel Model for Securing Mobile-Based Systems against DDoS Attacks in Cloud Computing Environment', in *International Journal of Interactive Mobile Technologies*, vol. 13/1, pp. 85–98. DOI: 10.3991/ijim.v13i01.9900.

Echeverria, J. 2017. *El Arte de innovar. Naturalezas, lenguajes y sociedades*, Madrid: Plaza y Valdés.

Elster, J. 1983. *Explaining Technical Change*, Cambridge: Cambridge University Press.

Elton, M. C. J. and Carey, J. 2013. 'The Prehistory of the Internet and the Traces in the Present: Implications for Defining the Field', in W. H. Dutton (ed.), *The Oxford Handbook of Internet Studies*, Oxford: Oxford University Press, pp. 27–47.

Eubanks, V. 2018. *Automating Inequality: How High-Tech Tools Profile, Police, and Punish the Poor*, New York: St. Martin's Press.

Fleming, L. and Sorenson, O. 2004. 'Science as a Map in Technological Search', in *Strategic Management Journal*, vol. 25/8/9, pp. 909–928.

Floridi, L. 2011. *Philosophy of Information*, Oxford: Oxford University Press.

Floridi, L. 2014. *The Fourth Revolution – How the Infosphere is Reshaping Human Reality*, Oxford: Oxford University Press.

Gonzalez, W. J. 1997. 'Progreso científico e innovación tecnológica: La 'Tecnociencia' y el problema de las relaciones entre Filosofía de la Ciencia y Filosofía de la Tecnología', in *Arbor*, vol. 157/620, pp. 261–283.

Gonzalez, W. J. 2005. 'The Philosophical Approach to Science, Technology and Society', in W. J. Gonzalez (ed.), *Science, Technology and Society: A Philosophical Perspective*, A Coruña: Netbiblo, pp. 3–49.

Gonzalez, W. J. 2013a. 'The Roles of Scientific Creativity and Technological Innovation in the Context of Complexity of Science', in W. J. Gonzalez (ed.), *Creativity, Innovation, and Complexity in Science*, A Coruña: Netbiblo, pp. 11–40.

Gonzalez, W. J. 2013b. 'The Sciences of Design as Sciences of Complexity: The Dynamic Trait', in H. Andersen, D. Dieks, W. J. Gonzalez, Th. Uebel and G. Wheeler (eds.), *New Challenges to Philosophy of Science*, Dordrecht: Springer, pp. 299–311.

Gonzalez, W. J. 2014. 'The Evolution of Lakatos's Repercussion on the Methodology of Economics', in *HOPOS: The Journal of the International Society for the History of Philosophy of Science*, vol. 4/1, pp. 1–25.

Gonzalez, W. J. 2015a. 'On the Role of Values in the Configuration of Technology: From Axiology to Ethics', in W. J. Gonzalez (ed.), *New Perspectives on Technology, Values, and Ethics: Theoretical and Practical*, Boston Studies in the Philosophy and History of Science, Dordrecht: Springer, pp. 3–27.

Gonzalez, W. J. 2015b. *Philosophico-Methodological Analysis of Prediction and Its Role in Economics*, Dordrecht: Springer.

Gonzalez, W. J. 2016. 'Rethinking the Limits of Science: From the Difficulties to the Frontiers to the Concern about the Confines', in W. J. Gonzalez (ed.),

The Limits of Science: An Analysis from 'Barriers' to 'Confines', Poznan Studies in the Philosophy of the Sciences and the Humanities, Leiden: Brill-Rodopi, pp. 3–30.

Gonzalez, W. J. 2017. 'From Intelligence to Rationality of Minds and Machines in Contemporary Society: The Sciences of Design and the Role of Information', in *Minds and Machines*, vol. 27/3, pp. 397–424. DOI: 10.1007/s11023-017-9439-0.

Gonzalez, W. J. 2018a. 'Internet en su vertiente científica: Predicción y prescripción ante la complejidad', in *Artefactos: Revista de Estudios sobre Ciencia y Tecnología*, vol. 7/2, 2nd period, pp. 75–97. DOI: 10.14201/art2018717597.

Gonzalez, W. J. 2018b. 'Complejidad dinámica en Internet como plataforma de información y comunicación: Análisis filosófico desde la perspectiva de Ciencias de Diseño y el papel de la predicción', in *Informação e Sociedade: Estudos*, vol. 28/1, pp. 155–168.

Gonzalez, W. J. 2019. 'Internet y Economía: Análisis de una relación multivariada en el contexto de la complejidad', in *Energeia: Revista internacional de Filosofía y Epistemología de las Ciencias Económicas*, vol. 6/6, pp. 11–36. Available at https://abfcfc9a-c7ef-4730-b66e-0a415ef434c0.filesusr.com/ugd/e46a96_b400af5a739e4310a31b7e952244745d.pdf Accessed at 1.4.2020.

Gonzalez, W. J. and Arrojo, M. J. 2019. 'Complexity in the Sciences of the Internet and its Relation to Communication Sciences', in *Empedocles: European Journal for the Philosophy of Communication*, vol. 10/1, pp. 15–33. DOI: 10.1386/ejpc.10.1.15_1.

Gonzalez, W. J. 2020a. 'Electronic Economy, Internet and Business Legitimacy', in J. D. Rendtorff (ed.), *Handbook of Business Legitimacy: Responsibility, Ethics and Society*, Dordrecht: Springer, pp. 1327–1347. DOI: 10.1007/978-3-319-68845-9_84-1.

Gonzalez, W. J. 2020b. 'La dimensión social de Internet: Análisis filosófico-metodológico desde la complejidad', in *Artefactos: Revista de Estudios de la Ciencia y la Tecnología*, vol. 9/1, 2 nd period, pp. 101–129. DOI: 10.14201/art2020101129. Available at https://revistas.usal.es/index.php/artefactos/article/view/art2020101129 Accessed at 27.4.2020.

Gonzalez, W. J. 2020c. 'The Internet at the Service of Society: Business Ethics, Rationality, and Responsibility', in *Éndoxa*, vol. 46, pp. 383–412. Available at http://revistas.uned.es/index.php/endoxa/article/view/28029/pdf

Gonzalez, W. J. 2021. 'Semantics of Science and Theory of Reference: An Analysis of the Role of Language in Basic Science and Applied Science', in W. J. Gonzalez (ed.), *Language and Scientific Research*, Cham: Palgrave Macmillan, pp. 41–91. DOI: 10.1007/978-3-030-60537-7_2.

Gonzalez, W. J. 2022. 'Scientific Side of the Future of the Internet as a Complex System. The Role Prediction and Prescription of Applied Sciences', in W. J. Gonzalez (ed), *Current Trends in Philosophy of Science. A Prospective for the Near Future*, Synthese Library, Cham: Springer, pp. 103–104.

Gonzalez, W. J. 2023a. 'The Internet as a Complex System Articulated in Layers: Present Status and Possible Future', in W. J. Gonzalez (ed.), *The Internet and Philosophy of Science*, Routledge Studies in the Philosophy of Science, London: Routledge, pp. 35–70.

Gonzalez, W. J. 2023b. 'Biology and the Internet: Fake News and Covid-19', in W. J. Gonzalez (ed.), *The Internet and Philosophy of Science*, London: Routledge, pp. 195–207.

Greenstein, S. 2015. *How the Internet Became Commercial. Innovation, Privatization, and the Birth of a New Network*, Princeton, NJ: Princeton University Press.

Guilló, M. 2015, 'Futures of Participation and Civic Engagement within Virtual Experiments', in J. Winter and R. Ono (eds.), *The Future Internet: Alternative Visions*, Dordrecht: Springer, pp. 41–57.

Hall, W., Hendler, J. and Staab, S. 2016. 'A Manifesto for Web Science @10', December 2016, pp. 1–4. Available at http://www.webscience.org/manifesto Accessed at 16.5.2018.

Hands, D. Wade. 1991. 'Reply to Hamminga and Mäki', in N. de Marchi and M. Blaug (eds.), *Appraising Economic Theories*, Aldershot: Edward Elgar, pp. 91–102.

Hendler, J. and Golbeck, J. 2008. 'Metcalfe's Law, Web 2.0, and the Semantic Web', in *Journal Web Semantics: Science, Services and Agents on the World Wide Web*, vol. 6/1, pp. 14–20.

Hendler, J. and Hall, W. 2016. 'Science of the World Wide Web', in *Science*, vol. 354/6313, pp. 703–704.

Hendler, J. 2023. 'The Future of the Web', in W. J. Gonzalez (ed), *The Internet and Philosophy of Science*, Routledge Studies in the Philosophy of Science, London: Routledge. pp. 71–83.

Henten, A. and Tadayoni, R. 2008. 'The Impact of the Internet on Media Technology, Platforms and Innovation', in L. Küng, R. G. Picard and R. Towse (eds.), *The Internet and the Mass Media*, London: SAGE, pp. 45–64.

Humphreys, P. 2016. *Emergence: A Philosophical Account*, Oxford: Oxford University Press.

Isaacson, W. 2011. *Steve Jobs*, New York: Simon and Schuster.

Kaufman, J. C. and Sternberg, R. J. (eds.). 2010. *The Cambridge Handbook of Creativity*, Cambridge: Cambridge University Press.

Leiner, B. M., Cerf, V. G., Clark, D. D., Kahn, R. E., Kleinrock, L., Lynchn, D. C., Postel, J., Roberts, L. G. and Wolff, S. 2009. 'A Brief History of the Internet', in *ACM SIGCOMM Computer Communication Review*, vol. 39/5, pp. 22–31.

Liu, J., Ahmed, E., Shiraz, M., Gani, A., Buyya, R. and Qureshi, A. 2015. 'Application Partitioning Algorithms in Mobile Cloud Computing: Taxonomy, Review and Future Directions', in *Journal of Network and Computer Applications*, vol. 48, p. 101. DOI: 10.1016/j.jnca.2014.09.009.

Lynch, M. P. 2016. *The Internet of Us: Knowing More and Understanding Less in the Age of Big Data*, New York: Liveright/W. W. Norton.

Mohapatra, S., Paikaray, J. and Samal, N. 2015. 'Future Trends in Cloud Computing and Big Data', in *Journal of Computer Sciences and Applications*, vol. 3/6, pp. 137–142. DOI: 10.12691/jcsa-3-6-6.

Murray, J. K., Studer, J. A., Daly, S. R., McKilligan, S. and Seifert, C. M. 2019. 'Designing by Taking Perspectives: How Engineers Explore Problems', in *Journal of Engineering Education*, vol. 108/2, pp. 248–275. DOI: 10.1002/jee.20263.

Niiniluoto, I. 1994. 'Nature, Man, and Technology – Remarks on Sustainable Development', in L. Heininen (ed.), *The Changing Circumpolar North: Opportunities for Academic Development*, Rovaniemi: Arctic Centre Publications, n. 6, pp. 73–87.

Novet, J., 2019. 'Amazon Web Services Reports 45 Percent Jump in Revenue in the Fourth Quarter', in *CNBC*, January 31, extended on February 1, 2019.

Available at https://www.cnbc.com/2019/01/31/aws-earnings-q4-2018.html Accessed at 25.8.2019.

Ordoñez, J. 1992. 'Los mecanismos de la innovación: La invención y los sistemas de patentes', *Arbor*, vol. 142, n. 558-559-560, pp. 253–270.

O'Reilly, T. 2005. 'What is Web 2.0? Design Patterns and Business Models for the Next Generation of Software', Available at http://www.oreilly.com/pub/a/web2/archive/what-is-web-20.html Accessed at 14.7.2022.

Park, J. H. 2019. 'Advances in Future Internet and the Industrial Internet of Things', in *Symmetry*, vol. 11/2, pp. 1–4. DOI: 10.3390/sym11020244.

Ramadani, V. and Gerguri, S. 2011. 'Innovations: Principles and Strategies', in *Strategic Change*, vol. 20/3–4, pp. 101–110.

Rescher, N. 1992. 'Our Science as *our* Science', in N. Rescher (ed.), *A System of Pragmatic Idealism. Vol. I: Human Knowledge in Idealistic Perspective*, Princeton, NJ: Princeton University Press, pp. 110–125.

Rescher, N. 1999. *Razón y valores en la Era científico-tecnológica*, Barcelona: Paidós.

Rescher, N. 2003. *Sensible Decisions. Issues of Rational Decision in Personal Choice and Public Policy*, Lanham: Rowman and Littlefield.

Schultze, S. J. and Whitt, R. S. 2016. 'Internet as a Complex Layered System', in J. M. Bauer and M. Latzer (eds.), *Handbook on the Economics of the Internet*, Cheltenham: Edward Elgar, pp. 55–71.

Shneiderman, B. 2019. 'Creativity and Collaboration: Revisiting cybernetic serendipity', in *Proceedings of the National Academy of Sciences*, vol. 116/6, pp. 1837–1843.

The Economist 2018. 'More Knock-on than Network. How the Internet Lost Its Decentralised Innocence', *Special Report: Fixing the Internet*, vol. 427/9098, June 30th, 2018, pp. 5–6.

The Economist 2019a. 'Send in the Clouds. Microsoft's Transformation Require a Chance of Culture', Section *Bartley*, July 6, p. 51.

The Economist 2019b. 'Digital Marketplaces. How to Beat Bezos', Section *Business*, August 3, pp. 53–54.

The Economist 2019c. 'Information Technology. The Digital Assembly Line', Section *Business*, September 7, p. 54.

The Economist 2019d. 'Ubiquitous Computing. Chips with Everything', Section *Technology Quarterly: The Internet of Things*, September 14, pp. 3–4.

The Economist 2019e. 'Big Tech and Antitrust. How to Dismantle a Monopoly', Section *Business*, October 26, pp. 57–59.

The Economist 2021a. 'The Online Boom. The Future of Global E-Commerce', Section *Leaders*, January 2, p. 10.

The Economist 2021b. 'Cryptocurrencies. If You Can't Beat Them', Section *Leaders*, January 9, p. 9.

The Economist 2021c. 'Freedom of Speech. Inconvenient Truths', Section *International*, February 13, pp. 49–50.

The Economist 2021d. 'Facebook and Donald Trump. Speechless', Section *United States*, May 8, pp. 36–37.

The Economist 2021e. 'Automating Programming. The Software Engineers', Section *Science and Technology*, July 10, pp. 67–68.

Timmers, P. 1999. *Electronic Commerce: Strategies and Models for Business-to-Business Trading*, Chichester: J. Wiley and Sons.

Tiropanis, T., Hall, W., Crowcroft, J., Contractor, N. and Tassiulas, L. 2015. 'Network Science, Web Science, and Internet Science', in *Communications of ACM*, vol. 58/8, pp. 76–82.

Varghese, B. and Buyya, R. 2018. 'Next Generation Cloud Computing: New Trends and Research Directions', in *Future Generation of Computer Systems*, vol. 79, pp. 849–861. DOI: 10.1016/j.future.2017.09.020.

Vasuki, K., Rajeswari, K. and Prabakaran, M. 2018. 'A Survey of Current Research and Future Directions Using Cloud-Based Big Data Analytics', in *International Research Journal of Engineering and Technology*, vol. 5/5, pp. 3841–3844.

von Hippel, E. 2005. *Democratizing Innovation*, Cambridge, MA: The MIT Press.

Wachter-Boettcher, S. 2017. *Technically Wrong: Sexist Apps, Biased Algorithms, and Other Threats of Toxic Tech*, New York: W. W. Norton.

Weber, B. 2016. 'Bitcoin and the Legitimacy Crisis of Money', in *Cambridge Journal of Economics*, vol. 40, pp. 17–41. DOI: 10.1093/cje/beu067.

Winter, J. and Ono, R. (eds.). 2015. *The Future Internet: Alternative Visions*, Dordrecht: Springer.

Yap, K. L., Chong, Y. W. and Liu, W. 2020, 'Enhance Handover Mechanism Using Mobile Prediction in Wireless Networks', in *PLoS One*, vol. 15/1, pp. 1–31. DOI: 10.371/journal.phone.0227982.

Part I

Configuration of the Internet and Its Future

2 The Internet as a Complex System Articulated in Layers

Present Status and Possible Future[1]

Wenceslao J. Gonzalez

2.1 The General Framework of the Complexity of the Network of Networks

A central feature of the Internet, understood in a broad sense, is that it is articulated in layers,[2] due to mainly structural and dynamic reasons, which give rise to diverse tasks according to each layer. These layers configure a unique whole, which is a complex system of network of networks and includes three main layers: (a) the technological infrastructure or the Internet in the strict sense, which is the basis on which the whole system was built; (b) the Web,[3] which relies on this support to perform a plethora of tasks; and (c) the cloud computing, along with apps and the mobile Internet. The aim here is to analyze this complex system from the layers that compose it, looking at its current status and its possible future. For this philosophico-methodological analysis, the starting point is the existence of a complexity that has a multivariate origin. Among the elements of its complexity are, at least, the following:

(1) The articulation of the layers involves a complexity of *origin*, insofar as there is a triple interaction of sources: the scientific side, the technological facet and the social dimension.[4] These sources affect the development and operation of the network of networks, so they are commonly intertwined at the service of society.[5] (2) Along with the complexity of origin, there are three forms of complexity in the *configuration* of this system: (i) structural, for the number and variety of components and the kind of arrangements involved, due to the various types of relationships between them; (ii) dynamic, for the factors involved in change over time, which are usually considered in terms of evolution, but which present more radical variations that can be considered in terms of historicity;[6] and (iii) pragmatic, because of the solutions (scientific, technological and social) to be proposed in increasingly varied contexts of use, which leads to new phenomena and specific situational circumstances.[7]

(3) Within the *articulation itself* of the complex system of the network of networks we can distinguish epistemological and ontological elements, which are available in the structural, dynamic and pragmatic factors.

DOI: 10.4324/9781003250470-3

Furthermore, they can exhibit various modes of complexity. Among the epistemic modes are descriptive, generative and computational ways of complexity, while the ontological modes include compositional (constitutional and taxonomical), genuinely structural (organizational and hierarchical) and functional (operational and nomic) ways of complexity.[8] (4) There are *internal and external features* in the origin of each of layer, in their components of their configuration (structural, dynamic and pragmatic) and in their elements of complexity (epistemological and ontological) present. In addition, these internal and external features are in constant interaction.

This last aspect of complexity, which highlights the presence of dualities as internal and external of undoubted philosophico-methodological interest, is the focus of the present analysis. The object of study is both the current characteristics and the possibilities for the near future of the three main layers of the network of networks. Thus, based on the existence of structural, dynamic and pragmatic criteria of complexity, the analysis follows three steps: first, the layer of the Internet *sensu stricto* as a network; second, the layer of the Web; and third, the layer of the cloud computing, practical applications (apps) and the 'mobile Internet'.

Regarding the complex system itself, it is important to point out that 'these three layers or tiers are conceived here as successive in dynamic terms, but not necessarily as dependent in terms of their structure. This involves that there is no intrinsic dependence – technological or otherwise – of the third level on the second level. Rather, the articulation is of this type: a) the Web is developed on top of the network (including mobile Internet) in order to have navigation and content; b) the cloud is built on top of the network (including mobile Internet) to provide resources for the storage of information and computing; and c) the cloud is later than the Web in terms of its composition and has interdependencies with it, but is not necessarily built on top of it' (Gonzalez 2022, p. 106).

2.2 The Layer of the Internet *sensu stricto* as a Network

To characterize the Internet *sensu stricto*, a central feature is its artificial ontology, through which data are sent, which are the basis of information and knowledge for those who receive the message. The Internet is a network of *physical data connections* between computers and other devices. 'The Internet is a *packet-switched* data network, meaning that messages sent over it are broken into *packets*, small chunks of data, that are sent separately over the network and reassembled into a complete message again at the other end. The format of the packets follows a standard known as *Internet Protocol* (IP) and includes an *IP address* in each packet that specifies the packet's destination, so that it can be routed correctly across the network' (Newman 2018, p. 15).

Along with the Internet Protocol, the *Transfer Control Protocol* (TCP) was established. TCP was designed to provide a standard for the reliable transmission of data between machines (i.e., the machines where the Internet Protocol operates). 'The TCP software takes the responsibility of breaking the data into packets, numbering the packets to detect losses and reordering, retransmitting lost packets until they eventually get through, and delivering the data in order to the application. This service is often much easier to utilize than the basic Internet communication service' (Clark 2018, pp. 10–11).

Through the artificial ontology of the network and the design of its protocols for the operation of this layer, we have key elements for the analysis of complexity of the Internet in the strict sense, following the internal-external perspectives. The *internal* viewpoint in the philosophico-methodological characterization of this layer – the technological infrastructure of the system and its services – can lead to a first feature: It is a human undertaking of a teleological character within an artificial setting and a social dimension. Thus, the internal configuration of the Internet as a network can be thought of as an expression of human activity, based on a scientific side that is goal-oriented and a technological facet.[9]

Additional traits of the activity of this layer are the following: (a) the network has a *main purpose* for its operation, which is connectivity between machines, primarily for information and communication (and it has progressively allowed more tasks over the years); (b) the *means* to make that operation of information and communication viable requires elements such as the 'Internet Protocol' (IP);[10] and (c) the *outcome* is thought to be the result of an open dynamic, which goes from part to part or from end to end, instead of making explicit something that is underlying or implicit in the network.

According to these internal traits of the activity of this layer, IP technology has a number of characteristics: (i) there is a separation between the technology of the network, as such, and the services, which have increased in number and variety; (ii) it has an internal end-to-end architecture, where there is an interrelation between the core and the edge of the network; (iii) IP technology has the trait of scalability; and (iv) it has a distributed design and a decentralized control,[11] which have contributed to its development.

IP technology is designed to facilitate the *viability* of the infrastructure platform and the Internet services, such as those related to communication, which are supported by the technological infrastructure of the network. However, 'it is generally recognized that the current approach of using the IP address both as a locator and as an identifier was a poor design choice. (...). In today's Internet [2018], dealing with mobility is complicated by the fact that the IP address is used both for forwarding and for end-node identity' (Clark 2018, p. 93).

Nevertheless, the Internet Protocol (IP) promotes *scalability*, which is the ability of the system to adapt to increasing demands. Scalability is one of the design principles of the IP, which includes that 'the network can scale up and new users and components can be added to the network in a simple and efficient way' (Henten and Tadayoni 2008, p. 52). In addition to scalability, the IP platform contributes to a *distributed design* and a *decentralized control*, which improves the conditions for the development of services as well as for other kinds of innovations, such as those related to new businesses.[12]

For the future, an important aspect of the Internet is the characteristic of scalability, in addition to issues of quality of service and security of this complex system. Moreover, it is assumed that this complex system has to be both robust and flexible. One of the barriers that we have for further scalability is the shortage of addressing space in the current protocols available for years in this complex system. In this regard, it seems that 'the shortage of addressing space is a problem especially for developing countries' (Henten and Tadayoni 2008, p. 53). This problem of a shortage of IP addresses is handled in the new version of the Internet Protocol.

Yet, the dominant Internet internal configuration, which is based on the client-server solution, is not *eo ipso* the most efficient for the organization of content distribution. The alternative is *Peer-to-Peer* (P2P) architectures, which are developed mainly for content sharing. This kind of networking is used in the media (music, film industry, etc.), where the peers communicate directly with each other and the IP terminals perform as clients and servers. 'P2P technology was actually one of the basic technology concepts of the Internet. It was described in the very first request for comments (RFC) from computers on the ARPANET' (Henten and Tadayoni 2008, p. 54). Although the Internet has been dominated by client-server technology for many years, it has allowed for P2P networking solutions, which have been used in many applications.[13]

Modularity is another feature of the internal configuration of the network to achieve a good level of operation of the Internet in the strict sense (effectiveness or, better, efficiency).[14] Consequently, the functional tasks are divided and assigned to different action protocols. Thus, the physical component deals with the electrical signals that are transmitted; the transport section deals with how the data packets are routed though the network to their correct destinations; and the application element controls how these packets are used to serve the users of the network.[15]

Socially, the Internet has expanded exponentially from its initial narrow institutional mission to its current dimension, which encompasses more than half of the world's population.[16] If the *external* perspective of the social dimension is considered, then the configuration of the layer of the network is ordinarily perceived by the user through the degree of proximity or immediate usefulness of the technology involved. *De facto*, there are two different kinds of technologies working in the network: (i) *backbone technology*, which is the background or ultimate support of the network of networks and which is supplied by the *Internet Infrastructure Providers*[17] and (ii) the technology of the *Internet Service Providers*,[18] which is offered by companies to users (organizations, groups or individuals), commonly through some kind of contract.[19]

It seems clear that the backbone technology is far from users, insofar as the users (mainly individuals, groups or small organizations) do not usually have direct access to this structural level. It is the core of the network and 'contains the trunk lines that provide long-distance high-bandwidth data transport across the globe, along with the high-performance routers and switching centers that link the trunk lines together. The trunk lines are the highways of the Internet, built with the fastest fiber optics connections available (and improving all the time)' (Newman 2018, p. 15).

Meanwhile, the Internet Services Providers is closer to the users, because it is the ordinary nexus of most users (individuals, groups or small organizations). The ISP contract with the Internet Infrastructure Providers or Network Backbone Providers for connection to the backbone technology, and either they resell the product or provide the connection to end users, who are the ultimate consumers of Internet bandwidth (businesses, academic institutions, people in their homes, etc.).[20]

Certainly, the priority of the designer of the network or those who take care of its proper functioning is different than that of the users of the technological platform. This internal-external difference in perspective can be seen when the 'users of the Internet tend to use the term *Internet* to describe the totality of the experience, focusing on the applications, but to a network engineer the Internet is a packet transport service provided by one set of entities (Internet service providers, or ISP); the applications run *on top of* that service. Anyone with the skills and inclination can develop a new application for the Internet' (Clark 2018, pp. 5–6).

As for the question of which is the last external instance that controls the Internet, the answer is not easy.[21] As a technological platform, 'the system organizes itself by the combined actions of many local and essentially autonomous computer systems' (Newman 2018, p. 17). A large amount of *autonomy* is a remarkable feature of the Internet in the strict sense. The *Border Gateway Protocol* (BGP) 'is designed in such a way that if new nodes or edges are added to the network or old ones disappear, either permanently or temporarily, routers will take note and adjust their routing policy appropriately. There is a certain amount of human oversight involved, to make sure the system keeps running smoothly, but no 'Internet government' is needed for to steer things from on high' (Newman 2018, p. 17).

2.2.1 *The Network and Main Factors Involved for Its Future*

Initially, the Internet – in the strict sense – is associated with an ontology of the artificial, which entails 'a set of nodes connected by links' (Bagnoli et al. 2019, p. 1).[22] This idea of a network is certainly broad. Its design is based on epistemological contents of scientific creativity and technological innovation. The design includes internal and external features, which address the past, present and future of this network. All together involve structural, dynamic and pragmatic aspects of complexity in this layer.

Each of the main factors related to the configuration of the future of the Internet as an artificial network has to consider the prediction, be it in the short, middle or long run. Since this infrastructure of the system considered is technological, we should think about the three types of predictions involved in the technology. First, the predictions of the sciences of the Internet (mainly, the Network science and the Internet science),[23] which have to offer scientific knowledge regarding the possible future in the face of the problems raised by this technological platform.[24] Second, predictions regarding the viability of the devices or technological elements and their cost-benefit ratio, which influence the decision to carry out or not this new technological component on this layer of the network. Third, predictions about internal and external values (especially those affecting technological ends and means), including consideration of endogenous and exogenous ethical values. This involves anticipating the acceptability or otherwise of technological objectives, means and outcomes, but also establishing priorities, where appropriate.[25]

These predictions have to consider, on the one hand, what the Internet in the strict sense has in common with other networks; and, on the other hand, what this network has that is singular or unique. What this network has in common or shares with other networks is 'how a structured network arises and grows, how it can be crossed, cut or made more robust, it can be explored and measured, and how the processes are affected by the network structure. However, the Internet should not be seen only at the interconnection level' (Bagnoli et al. 2019, p. 1). Consequently, it is not simply an inter-network that supports other networks, but a *unique* inter-network.

Among the factors of this uniqueness – which make for a complex system and generate difficulties to the prediction, due to the complexity – are that this inter-network combines three main philosophico-methodological components:[26] (a) the epistemological-methodological, through the *protocols* related to information and the processes to its transmission over long distances and with great speed and high level of reliability; (b) the ontological-methodological, through a *physical medium* – a sophisticated technology – that is able to send this information by adequate channels to reach the intended recipients; and (c) the ontological-axiological, through the *activity of human agents* who, within a contemporary social environment, produce information for very diverse *purposes* according to a set of accepted values. All three factors can be related to predicting the possible future and prescribing patterns of action, which is the task of applied science.

From a scientific viewpoint, the characteristics that make this network that we call 'Internet' unique are studied mainly by two sciences, which are primarily in the field of the artificial, although they have a clear inter-disciplinary component,[27] and which pay great attention to the future of the Internet. These are Network science and Internet science, which make predictions and prescriptions, especially when they provide scientific content – solutions to specific problems – for the technological facet of this complex system.

Network science deals with what this *inter-networking* will involve with other networks. Thus, it pays special attention to the physical infrastructure, which allows a high level of communication between machines that can be across short, middle or long distances. Meanwhile, Internet science is more focused on *content* than on the support itself. Thus, 'Internet science can be considered a research field that investigates the aspect of communication among actors and resources that can shape the information' (Bagnoli et al. 2019, p. 1), which is the content transmitted on the Internet.

Besides the internal perspective in scientific-technological terms, there is an external viewpoint, which is evident in terms of the social dimension. This is where the Internet – in the strict sense – can have problems in the future in several directions. Among them are the following: (1) the loss of neutrality, leading to the development of a 'digital sovereignty' by countries (such as Russia or China),[28] where the network ends up being de facto controlled in those territories; (2) the creation of a two-speed Internet, whether its services are paid for or not; and (3) the sectorialization of the network, to create branches such as commercial, educational, health, financial, etc.

2.2.2 The Internet as a Scale-Free Network and Its Future

Ontologically, the Internet *sensu stricto* is primarily configured in the realm of the artificial as technological infrastructure, which is the result of social action, both collaborative and competitive. This complex artifact is a 'physical network where the nodes are routers and the links are physical connections' (Barabási 2013, p. 1). This physical network has developed exponentially, but – for the future – it requires scientific contribution to propose new solutions to the technological facet and a new impulse to the social dimension.

Meanwhile, also within the realm of the artificial, 'the Web is an information network, in which the nodes are documents (at the time of writing over one trillion of them), connected by links' (Barabási 2013, p. 1).

The Network science seeks to characterize these two large networks, to explain their configuration, to give reasons for their development and to evaluate their strengths and weaknesses. Using its own epistemological and methodological apparatus, network science studies emerging networks and considers how they change. These changes in the two initial networks – the Internet and the Web – have enabled the third main layer (the cloud computing, apps and the mobile Internet).

Conceived in the strict sense, 'the Internet is a scale-free network,[29] as are many online communities,[30] as well as networks in the natural world, such as the metabolic network of organisms (...). Such networks are in effect held together by a small number of highly connected hubs' (Barabási 2013, pp. 1–2). This *scale-free* character of the Internet as a complex network is very important for the future, since it entails the *strength of resilience* in case of random error and, at the same time, it is a *relevant weak point* in case of a serious attack on the network.[31] In this regard, there are two relevant aspects, which are initially structural, but which have clear consequences for the dynamics of this complex system.

(i) Resilience against random error or decay is one important property of scale-free networks. Thus, 'if one removes nodes from a network at random, then most types of network will eventually fragment into a set of smaller or individual nodes. But a scale-free network remains robust against random decay; it will shrink but not fall apart' (Barabási 2013, p. 2). (ii) The *Achilles' heel* of the scale-free network is 'that, in the event of a targeted attack, where the most connected nodes are deliberately removed first, then the network will be destroyed very quickly' (Barabási 2013, p. 2). This major vulnerability of the Internet as a network requires applied science: predicting when and how it may occur, and then prescribing guidelines to avoid or mitigate that possibility. In addition, the appropriate technology would be needed to fix the problems.[32]

2.2.3 *The Case of the Internet of Things*

Another ontological element, which is also primarily of artificial kind, is the Internet of Things (IoT), 'previously called sensor and actuator networks' (Clark 2018, p. 113).[33] On the one hand, the IoT is connected to the technological infrastructure of the Internet *sensu stricto* and has had a real impulse in recent years. But, regarding the future, 'some advocates of the IoT assert that the current Internet protocols will not serve these sorts of devices well', although Clark maintains that 'almost any device can be programmed to send an Internet packet' (Clark 2018, p. 113). On the other hand, he sees a challenge for the future: 'the IoT environment raises issues related to management and configuration that the current Internet architecture does not address at all, and some devices will indeed use a network technology that is not compatible with Internet packets' (Clark 2018, p. 113).

Nonetheless, Clark considers that the future 'interconnection between the IoT network and the current Internet will happen in the application layer, not at a lower transport layer. IoT interconnection will not be implemented by a general overlay or interconnection architecture' (Clark 2018, p. 113).[34] Looking at this issue not from the scientific side and the technological facet but from the social dimension, the Internet of Things looks to the future when 'people and objects will be connected at any time, at any place, with anything and anyone and will utilize any network and services' (Park 2019, p. 1). In this regard, there might be a connection between the future of the Internet and the IoT, which can become a technology for manufacturing (i.e., an industry).

Considered from the external perspective, the *advance* of the Internet – in the broad sense – has had an impact on the IoT (in sensors and actuators), which can be seen in terms of scalability, security and availability. It can also be thought that, in order to achieve more progress in the IoT, many aspects from an internal point of view should be pondered, where scientific creativity and technological innovation interact. Among the aspects, three are particularly relevant for the future: (1) the communication networks working in terms of quality, (2) the processes seeking optimization of the resources for network operations and (3) privacy and security combined in the sending of data (Park 2019, pp. 2–3).

Thinking about the future of the IoT in terms of scalability, security and availability, heuristic algorithms can play a key role in predictive terms. This can be considered as one of the scientific contributions to the design of the IoT. But it happens that, when future proposals seek to optimize processes, one of the problems is the existence of computational complexity. Faced with these issues, scientific and technological designs related to the IoT can focus on the contributions of Artificial Intelligence. In this regard, in philosophical terms, machines are intelligent in the operational sense.[35] Thus, the machines 'are more efficient than humans in capturing and communicating data accurately and consistently' (Park 2019, p. 1).[36] But, when there is an operational failure, we are faced with a challenge, insofar as 'inefficient data transmission can cause significant personal and property loss' (Park 2019, p. 2).

Furthermore, the technological *fragmentation* of the IoT is another point of concern for the future, where internal and external elements are combined. This fragmentation is expressed in 'the coexistence of different standards, protocols and architectures that we will need to put in place in the future' (Park 2019, p. 4). If this is considered within the broader framework of complexity (structural, dynamic and pragmatic), which certainly has implications for the future, one of the priority issues of the Internet of Things is the choice between (a) hierarchical or centralized approaches and (b) distributed or peer-to-peer options.[37] They have their advantages and disadvantages:

From one side, the distributed approach is more scalable, but it requires special techniques for insuring the reachability of each node, the coverage of the whole domain, and the optimization of communications among different nodes. From the other side, a distributed approach implies the cooperation of the participating nodes. While this cooperation can be taken for granted when all sensors/devices belong to the same organization, it can break for heterogeneous devices or in the case of failure/hacking.

(Bagnoli et al. 2019, p. 3)

2.3 The Layer of the Web

Above the Internet layer in the strict sense – the technological platform that supports the network of networks – is the layer of the Web. It has grown intensely since it was made available to the general public on April 30, 1993.[38] There is a structural and dynamic nexus between these two ontological spheres (the Internet and the Web), which belong primarily to the field of the artificial, although both also have a social dimension,[39] which, in the case of the Web, is more obvious to the general public. The Web is an information network in which 'the nodes are web pages, containing text, pictures, or other information, and the edges are hyperlinks that allow us to navigate from page to page' (Newman 2018, p. 33).

At the same time as the design of the World Wide Web, there were similar information systems related to the Internet available competing for dominance, 'but the Web won the battle, largely because the inventors decided to give away for free the software (…) on which it was based—the Hypertext Markup Language (HTML) used to specify the appearance of pages and the Hypertext Transport Protocol (HTTP) used to transmit pages over the Internet. The Web's extraordinary rise is now a familiar part of history and most of us use its facilities at least occasionally, and in many times daily' (Newman 2018, p. 33).

By means of the new layer (webmail, social networks, video-streaming, over-the-top television, etc.), the complexity of the network of networks increased significantly. Moreover, a few years ago, the problems for the new generation of Internet to keep up with the growth of the Web were not few:

The Internet underlies the World Wide Web, but there are numerous challenges in ensuring it remains capable of supporting the Web. These include problems of scalability, guaranteeing high levels of performance, security, real-time adaptability, resilience and mobile communications. The next generation Internet will have to focus on developing solutions for these challenges.

(Shadbolt et al. 2013, p. 3)

What is at stake is a *universal architecture* for the Internet to be able to maintain a *single Web*. In this regard, the Internet brought novelty to computing and programming languages.[40] This novelty is a challenge for the future of the Web and computing in general.[41]

Over the years, the Web has grown through various kinds of distributed architecture, which have been directed towards a progressive *personalization*,[42] which seems to be the dominant theme for the future of the Web in the task of strengthening its social dimension.[43] Thus, according to Anne-Marie Kermarrec, 'the Web has become a user-centric platform where users post, share, annotate, comment and forward content be it text, videos, pictures, URLs, etc.' (Kermarrec 2013, p. 1). This change has consequences in terms of content and operations, due to variations in the use of the Web.

Kermarrec considers that 'this social dimension creates tremendous new opportunities for information exchange over the Internet, as exemplified by the surprising and exponential growth of social networks and collaborative platforms. Yet, niche content is sometimes difficult to retrieve using traditional search engines because they target the mass rather than the individual. Likewise, relieving users from useless notification is tricky in a world where there is so much information and so little of interest for each and every one of us. We argue that ultra-specific content could be retrieved (…) and notification be *personalized* to fit this new setting' (Kermarrec 2013, p. 1).

2.3.1 The Future of the Web as a Threefold Undertaking

It happens that, in a way somewhat similar to the case of the Internet *sensu stricto* (the technological infrastructure), the future of the Web has a scientific side, a technological facet and a social dimension. The challenge of the future affects all three, because it is a threefold undertaking. This challenge has several faces: (i) the demand for *continuity* in what has been the basis of the success of the Web and (ii) the need for *novelty* in two different ways: a longitudinal or horizontal novelty, which allows to enhance what already exists, and a transversal or vertical novelty, which generates a more intense dynamic of historicity and leads us to new aspects not contemplated until now in the Web. Commonly, in the challenge regarding the future of the Web (in the short, middle or long run), the focus is rarely on the scientific side, since the technological facet and social dimensions are highlighted.

To deal with the challenge of the future of the Web, according to the threefold undertaking, we should think of *three types of predictions*: first, a prediction made in scientific terms, primarily focused on the internal development, considering that its future is more an open process than an established point of arrival or a clear cut destination;[44] second, a technological prediction, which deals with the network support to the Web in order to make that prospect possible; and third, a social prediction, which has to do with the dynamics of the social dimension of the Web towards that possible outcome.[45]

As for the scientific side of the Web, which is the main focus of this paper, there are two completely different aspects involved, which should be distinguished in philosophico-methodological terms, because they modulate the type of prediction to be made. They are, first, the *use of the Web in science*, which was the initial focus of scientific research and, second, the *science of the Web*, which was developed as such later on. In the first case, a prediction is primarily made *on* the contents of the Web, whereas in the second case, a prediction is made predominantly *within* the Web as something to be developed in terms of scientific creativity (in addition to technological innovation and being aware of the social dimension).[46] This scientific creativity needs to extend the science of the Web to address ambivalences of the Web through new methods, which can include crossdisciplinarity in addition to interdisciplinarity, and with the support of research infrastructures.[47]

(I) Initially, it was the *use of the Web in science*, which has grown exponentially. Now, scientists and researchers cannot imagine their work without the World Wide Web.[48] This has multiple scientific expressions; among them are the following: (1) The Web as a *scientific instrument* in research (mainly for bibliographical information, which includes full references of articles or books, etc.); (2) the Web as a *support* to make known results already achieved, which are made accessible to the public, either on open access or by other means; and (3) the Web as a place where *multiple data* are found, due to the number of users and the diversity of tasks reflected, that can serve as a basis for new research (e.g., in the social sciences).[49]

(II) Afterwards, there is the *science of the Web*, which has become particularly visible since 2006,[50] when – by means of a 'scientification' of a series of procedures that were already available –[51] it became a body of organized knowledge to solve problems. One of the central aspects is that 'the Web is (...) a set of protocols and formalisms' (Shadbolt et al. 2013, p. 1). This can be interpreted as a science of the artificial that extends the possibilities of technological support,[52] which is also at the service of society.[53] Among the elements of this recent science is predicting the future of the Web – the possible trajectory of its protocols and formalisms to expand human possibilities and solve new problems.

When the focus is on the Web science, there are two main aspects for the future. First, to contribute directly to the development of the *Web as such*, using 'own methods and techniques' (Shadbolt et al. 2013, p. 5). This can be done – in my judgment – by maintaining in principle its status as primarily a science of design, within the sciences of the artificial,[54] but open to other scientific disciplines, commonly in an interdisciplinary undertaking.[55] Second, through the development of Web studies from the *emerging properties* of the Web, which generates new data and information, which should lead to new knowledge, both for Web science and for other scientific disciplines (economics, communication, information science, etc.).

In this second direction moves the creator of the first Web – Tim Berners-Lee – when he proposes the Web of Linked Data, thinking of the emergence of a Web of data. He expressly discusses issues that will affect the future of the Web and 'argues for the importance of the continued development of the Linked Data Web, and describes the use of linked open data as an important component of that' (Berners-Lee and O'Hara 2013, p. 1). In addition, Berners-Lee 'defends the Web as a read–write medium, and goes on to consider how the read–write Linked Data Web could be achieved' (Berners-Lee and O'Hara 2013, p. 1). What he wants is to link 'islands of data' and to provide a 'read-write' environment with people and machines able to interconnect and interoperate at the data, information and knowledge levels.[56]

2.3.2 *The Future of the Web as an Interaction of Internal and External Perspectives*

From an *internal* perspective, in order to anticipate the possible future of the Web, it seems necessary to start from its configuration as a complex system in a digital setting and, therefore, within an artificial realm. This configuration is initially artificial in its scientific design, which had the technological support that existed at the time (a NeXT computer used by Tim Berners-Lee)[57] and had a clear social dimension (first, for the CERN and, thereafter, for the world as a whole). This internal configuration involves structural, dynamic and pragmatic factors as well as epistemological and ontological elements, which include, at least, the following features of the Web.

(i) *Emergence* in epistemological and ontological terms, where 'complex structures emerge from apparently simple principles' (Shadbolt et al. 2013, p. 2).[58] These complex structures include the development of the infosphere through the blogosphere, the creation of tools such as Wikipedia, innovations introduced by social networks, etc.

(ii) *Connectivity* basically conceived as symmetry in the configuration, insofar as the distribution of pages – although some have many more links than others – 'looks largely the same at whatever scale you sample' (Shadbolt et al. 2013, p. 2).

(iii) A *near scale-free* internal configuration combined with a particular resilience when dealing with an attack.[59] It seems 'near scale-free' due to two features that are oriented in different directions: (a) if a majority of web pages are removed, it is usually possible to access them from one page to another by means of a path that remains available;[60] and (b) 'removing a relatively small number of the more highly connected items would lead to the disintegration of the network' (Shadbolt et al. 2013, p. 2).

(iv) *Great accessibility* as a network, because, in a limited number of steps, you can have access to the contents sought, which makes it possible to understand the sociological success and psychological acceptability of the Web.[61]

(v) *Sustainability* of the Web understood in a twofold way: (1) to maintain the structural and dynamic properties that have led to its operability and growth: openness, operational flexibility and versatility to adapt to new needs; and (2) to be able to face present and future challenges, such as fake news, cybercrime and attacks on privacy due to the misuse of personal or professional data and biased algorithms used by artificial agents. To the extent that the future Web will depend more on Artificial Intelligence,[62] sustainability means orienting AI towards a positive development – where scientific creativity and technological facets are at the service of the common good of society – and reducing the negative factors,[63] both those that already exist and future ones, those that are unacceptable from an ethical point of view and lack legitimacy for stakeholders.

But, from an *external* perspective, there is also a complex structure and dynamics that is detectable through the social dimension of the Web. This complexity involves a set of aspects, among which are those that are related to the starting point, the means that make it possible and the environment (social, cultural, economic, etc.) in which it moves. The main factors here are the following: (a) the *designers* who think about potential users of the Web and their needs and preferences, (b) the *agents* – individuals, groups and organizations – that are *users* and interact with the Web, and (c) the *regulations* of an international, national or regional nature that affect Web pages and the multitude of expressions from macro, meso and micro business initiatives.

Undoubtedly, both perspectives – internal and external – look to the possible future but do so in quite different ways from the viewpoint of the dynamics. This variety of options looks towards *historicity* in the changes. In this regard, in the dynamics of the Web, there is not only a continuity or a longitudinal or horizontal novelty – an evolution with adaptation to new digital environments – but also a vertical or transversal novelty, which represents a *revolution* (such as YouTube, over-the-top television, some contributions of the social networks, etc.).

Analyzing this deep change, it seems that first it was conceptual (which gave rise to new designs), then it was in the processes to be followed (the methodological component) and, finally, it took shape in the results achieved. This phenomenon has often happened in the history of the Web, from the initial design of March 12, 1989, to the current situation. *De facto*, there have been radical changes in aims, processes and results between Web 1.0 and the present version of the Web. The development itself reinforces that 'large amounts of the Web's content and structure are now created dynamically' (Shadbolt et al. 2013, p. 3).

As regards the features of continuity of the Web, looking from the *internal* perspective, 'there is work that seeks to ensure that the Web remains distributed yet stable, open and at the same time secure. The principles of universal access and non-proprietary, open formats underlie the Web's basic protocols. Changes to these formats or to the assumption of content accessibility could have far-reaching consequences that need to be understood' (Shadbolt, Hall, Hendler and Dutton 2013, p. 5). Meanwhile, looking from the *external* perspective, it is suggested that the organization of the web should remain as loose as possible, with obvious flexibility: 'The use of Web-based applications such as social media, online social networking, and wikis, for example, has facilitated peer production, crowd-sourcing, widespread network effects, new organizational forms and a general 'deformalization' of organizations' (Shadbolt, Hall, Hendler and Dutton 2013, p. 4).

On the novelty side – within the internal perspective and, especially, in what mainly concerns the scientific side – there is the role to be played by novel aims, new forms of heuristics and some developments of the Artificial Intelligence. To a large extent, the future of the Web is associated with new forms of *heuristics*, so 'as the Web grows, changes and expands, researches must seek novel ways to explore, navigate and search its content (…) the vast information spaces that are continuously emerging on the Web' (Shadbolt et al. 2013, p. 3). Heuristics belongs, in principle, to the internal perspective of the Web, so the future development of the Web depends on the internal dynamics that entail methodological advances, both in terms of search procedures – heuristics – and proper methods (i.e., processes of discovery with scientific support).

In addition to novel aims related to this layer and new forms of heuristics, the future of the Web now seems to depend on the development of Artificial Intelligence and finding an alternative 'to address problems that computers cannot yet solve' (Shadbolt et al. 2013, p. 4).[64] On the one hand, 'over the coming years, Web Science will focus closely on the emergence of Artificial Intelligence and its implications for the Web' (Berendt et al. 2020, p. 26).[65] On the other hand, 'although computers have advanced dramatically in many respects over the last 50 years, they still do not possess the basic conceptual intelligence or perceptual capabilities that most humans take for granted' (Shadbolt et al. 2013, p. 4).[66]

About the second issue, Luis von Ahn considers that 'the human brain is an extremely advanced processing unit that can solve problems that computers cannot yet solve. And because of the Web, we can consider humanity as an extremely advanced and large-scale distributed processing unit that can solve large-scale problems that computers cannot yet solve. Currently, we have a very parasitic relationship with computers' (von Ahn 2013, p. 3). In this regard, von Ahn has 'advocated a more symbiotic relationship, where humans solve some sub-problems, computers others, and the two are combined to address large-scale problems' (von Ahn 2013, p. 3). In this regard, von Ahn has 'advocated a more symbiotic relationship, where humans solve some sub-problems, computers others, and the two are combined to address large-scale problems' (von Ahn 2013, p. 3).

Pondered from the *external* perspective, the longitudinal or horizontal novelty with regard to governance of the future Web calls for increasing the developments in favor of blurring the state-societal boundaries. This trait is understood in terms of more openness based on individual and social responsibility,[67] which can 'support a move towards "open-book" governance, transparency and open data initiatives. These hold the promise of co-production and co-creation of government services' (Shadbolt et al. 2013, p. 4).

Moreover, due to the strong external perspective of the Web, in the future it can further enhance the social dimension in several directions. Besides the contents of interest for a broad public, the enhancement can be encompassed in terms of services and management. Among them might be the following: (1) health services, which will be increasingly digitized worldwide;[68] (2) personalized entertainment, which will expand the current audiovisual possibilities in fields such as music, television or cinema; (3) electronic commerce, which has had a special growth in the periods of lockdown because of Covid-19;[69] (4) new forms of online education, such as webinars; and (5) security issues for the users,[70] either individuals, groups or organizations, who want 'secure, trusted, and ubiquitous accesses to services' (Schönwälder et al. 2009, p. 30)[71] as well as a dynamic management.

Research is now particularly needed in how to deal with the negative aspects of the Web in terms of contents, services and management. This includes a role of moderation, especially regarding the *contents* in several directions. First, those related to the creation and dissemination of fake news, to be distributed by social networks (especially Facebook and Twitter), which affect issues as important as vaccines, the characteristics of Covid-19 or electoral periods. Second, to ensure that the rights of artistic creators, book authors or other cultural fields are respected.[72] Thirdly, to ensure that the content expressed – and maintained on the Web – is not an insult or expression that, within a citizen environment, could become a crime by undermining the fame or prestige of the persons, a threat to individuals or groups, an incitement to violence (sexual, political or social), etc. Moderation requires adequate means on the part of companies that have social networks and, where appropriate, regulations at the national or international level that prevent clear harm to individuals or communities.

2.3.3 Three Possible Scenarios for the Future of the Web

Overall, there may be a global image of the Web in the near future. James Hendler has considered three possible scenarios: (I) The dystopian future, (II) a more open Web, and (III) a position in between with features of both.[73] This analysis shows that there are elements of content, services and management. In this regard, the internal perspective of the development of the Web is combined with the external perspective of the factors of the social dimension that influence its future (companies, governments, organized groups, regulations, etc.). This external perspective has special weight in that possible future.

(I) The dystopian future is when negative trends spread, which undermine the openness and pluralism of the Web, such as (i) the decrease in competitiveness caused by the expansion of large companies in the sector (Google, Baidu, etc.), which can reach a clear domain of the Web (international, national, etc.); and (ii) control of information by governments, either within the country or through interference in other countries (e.g., in electoral processes), or well organized groups of quite different kinds (political, economic, social, etc.).

(II) A more open Web may come through free market values, democratic systems and criteria of transparency, which counterbalance the negative side of the Web. Among these positive elements are the Open Science, 'social machines' that perform an informative task (like Wikipedia) or the regulations for data protection, like the one approved by the European Union, which has been in place since 25 May 2018.[74] The proposal of the 'Internet Bill of Rights' in the United States is oriented in this direction and contemplates the following: (a) right to universal Web access, (b) right to net neutrality, (c) right to be free from warrantless metadata collection, (d) right to disclose amount, nature and dates of secret government data requests, (e) right to be fully informed of scope of data used, and (f) right to be informed when there is change of control over data.[75]

(III) A position in between the first and second option with features of both. For Hendler, the likelihood is that the future situation of Web will be somewhere in between the previous options and sharing features of both approaches. In this regard, the scale of the Web is enormous as is its underlying complexity (which has been emphasized here and in three directions: structural, dynamic and pragmatic). Thus, it is sensible to maintain that predicting how the future of the Web will be deployed is extremely difficult.[76] This third possibility is situated in a very wide spectrum, without having a guarantee of achieving what Simon liked to call 'satisficing,'[77] which is a good enough outcome, once either maximization or what Nicholas Rescher calls 'optimization'[78] have been ruled out.

Nonetheless, in addition to describing the possible future of the Web, within the framework of heuristic predictions based on (i) the variables known at the present time and (ii) the degree of control we have over them, it is possible to guide the possible future by offering prescriptions on how things should be. This is what Tim Berners-Lee has done through his *Contract for the Web*, presented by the World Wide Web Foundation in order to offer 'a global plan of action to make our online world safe and empowering for everyone' (World Wide Web Foundation, November 2019).

With *Contract for the Web*, Berners-Lee seeks to preserve the aims of bringing people together and freely available knowledge. The document sets out commitments in order *to guide* digital policy agendas. This involves three kinds of main agents of the external perspective of the Web: governments, companies and civil society (individuals, groups and organizations). The text is focused on what ought to be done by the agents. Thus, along with the specification of what each of kind of agent should preserve or achieve, there is an annex that includes a connection between human rights and the criteria that are part of the substance of the contract's message.

2.4 The Layer of the Cloud Computing, Practical Applications (Apps) and the 'Mobile Internet'

What configures the third main layer of the network of networks includes a set of elements: the cloud computing, practical applications (apps) and the mobile Internet. They are part of new forms of expression of the three types of complexity indicated: structural, dynamic and pragmatic. *De facto*, we are currently moving towards a richer ecosystem with multiple networks, which increases complexity:

> What is the role of that part that we call the global Internet? One role is individual access – the means by which people connect. Another role is peer-to-peer interaction – applications that do not depend on the cloud or centralized services.
>
> (Clark 2018, p. 307)

Both structurally and dynamically, multidirection can be considered one of the features of this third layer. This is the case insofar as the role of users (individuals, groups, organizations, governments, etc.) is in constant interaction with technological support of this layer in many directions. This social dimension of the cloud computing, apps and mobile Internet has been reinforced during the months of the Covid-19 pandemic. The competitiveness of companies related to this layer of the Internet – in the broad sense – is also another relevant feature, from which new possibilities for users are constantly emerging.

Historically, cloud computing started around fifteen years ago. The key reason was pragmatic, insofar as 'businesses began to outsource their webhosting, data centres, core computing systems and many applications to a few big providers, particularly the pioneer AWS, run by Amazon. The pandemic has shown just how critical the cloud has become. Many of the economy's main functions depend on it, including a wide range of e-commerce sites and applications that let you work from home. The scale of this activity is huge; approaching 10% of all technology spending is on the cloud. So are the sums of money being invested. Perhaps $40bn is being ploughed this year [2020] into data centres and other physical gear by AWS [Amazon Web Services] and others' (The Economist. 2020b, p. 13).

Based on operative needs, technological innovations and e-commerce demands have given a boost to cloud computing. Nowadays, there is intense competition between Azure, which Microsoft launched to the market four years after Amazon's AWS, and Alphabet's Google Cloud Platform (GCP). But there are also other important players in this field, such as Oracle and Alibaba Cloud. It happens that big technology companies have a frequent interrelationship between cloud computing and the mobile Internet. In this regard, the pandemic has served to highlight how the complexity of the system is accompanied by failures in operational performance. These weak spots are seen when there is an unusual phenomenon, which affects the structure or dynamics of this main third layer of the network of networks.[79]

2.4.1 The Cloud Computing and the Role of Prediction

Although the name "cloud" is assumed, David Clark has pointed out that 'the word *cloud* is a bit misleading – it tends to imply something amorphous and indefinite in form. Cloud computing, in its physical manifestation, is anything but indefinite in form – cloud computing platforms can be buildings the size of a football field, housing hundreds of thousands of processors, and drawing (and dissipating) megawatts of power' (Clark 2018, pp. 301–302). In addition, cloud computing belongs primarily to the third main layer of the Internet in the broad sense. But now there is something more in structural and dynamic terms, which add more complexity to the system.

Actually, cloud computing is related to other networks as well. In this regard, 'as the Internet ecosystem has expanded, the creation of the global experience no longer depends solely on a single interconnected Internet. Today, there are other global networks that are based on the same Internet technology but not directly interconnected with what we think of as the public Internet. These other networks have become part of the application development ecosystem, along with cloud computing, CDNs [content delivery networks], and the like. For example, cloud providers use these networks to reach their enterprise customers, thereby shielding that traffic from various attacks and fluctuations in performance' (Clark 2018, p. 307).

Placed in a context of complexity, we can consider the 'internal' configuration of the cloud computing and its 'external' relations to other layers of the Internet in the broad sense as well as other networks. The 'internal' setting is commonly associated with a platform for big data. Thus, there are two main tasks of the cloud with respect to big data: (1) a large capacity to store data (with the possibility of giving rise to information and knowledge) and (2) a huge computing capacity, which favors the use of content for multiple purposes.[80] In this regard, 'the cloud computing environment offers development, installation and implementation of software and data applications "as a service." Three multi-layered infrastructures namely, platform as a service (PaaS), software as a service (SaaS), and infrastructure as a service (IaaS), exist' (Vasuki et al. 2018, p. 3842).[81]

However, Azure considers these three (PaaS, SaaS and IaaS) as *services* rather than 'layers.' Moreover, this Microsoft's cloud adds a fourth service, which is 'service-less computing,' which overlaps with PaaS. In this respect, it builds a functionality for practical applications (apps), but without the need to spend time handling the servers and the corresponding infrastructure. This means that, in this case, the cloud offers a framework for operating according to specific functions or when something specific happens.[82]

Therefore, cloud computing relies on the Internet, but it also uses other networks that have been established by means of the same Internet technology. Following this 'external' side, cloud providers use these networks with some frequency to reach their business consumers, thus protecting that traffic from various attacks and performance fluctuations.[83] Security and reliability are then two advantages for the users. These characteristics lead to a higher degree of confidence, when making cyber security predictions, thinking about possible hacker attacks, as ordinary computers directly connected to the network of networks are more vulnerable than the cloud.

Efficiency, security and scalability (or elasticity) are three main aspects related to data of cloud computing and the Internet of Things.[84] Commonly, the cloud infrastructure is good enough for the tasks of storage and computing requirements of data analytics algorithms. *De facto*, 'the cost of storage has [been] considerably reduced with the advent of cloud-based solutions. In addition, the 'pay-as-you-go' model and the concept of commodity hardware allow effective and timely processing of large data, giving rise to the concept of 'big data as a service.' An example of one such platform is Google BigQuery, which provides real-time insights from big data in the cloud environment' (Vasuki et al. 2018, p. 3842).[85]

Although there are 'open issues like security, privacy and the lack of ownership and control' (Vasuki et al. 2018, p. 3844), which to a large extent are also present in other layers of the network of networks and have repercussions for the future, the cloud computing also has its own characteristics, both considered in its structure and dynamics and based on the content it stores. Thus, with respect to the cloud computing itself, there can be two types of predictions. On the one hand, predictions about this important layer of the network of networks and, on the other, predictions made using the content available in the cloud computing. These two types of predictions are complementary from a philosophico-methodological point of view. They can contribute to new functions of the cloud computing, in the first case, and to scientific, technological and social progress, in the second case:

(I) Predictions can be made regarding the *possible future of the cloud computing* in terms of various kinds of factors related to its configuration or its way of changing over time: (a) external factors, such as those affecting content (such as hackers), services (PaaS, SaaS and IaaS) or management (such as the presence of new companies capable of using data from various clouds, currently in competition)[86] from outside the layer; and (b) internal factors, such as new content storage capabilities, the introduction of new types of services,[87] or the limits of cloud management within the layer in the Internet conceived in the broad sense.[88]

Structural and dynamic factors are involved here and play a role in predicting the future of the cloud computing. On the one hand, the prediction has to look at the *structural* aspects of security and sustainability of future systems related to the cloud, taking into account scientific bases and possible new technological architectures. On the other, the new generation of the cloud has to meet the *dynamic* components of social demands that arise from users (individuals, groups or organizations). Actually, both types of factors are interconnected: 'new computing architectures are emerging. This change is impacting a number of societal and scientific areas' (Varghese and Buyya 2018, p. 849).

Predicting the new generation of cloud computing is not easy because of the structural and dynamic complexity. Among the epistemological and ontological elements to be considered for predictions are: (1) impact areas regarding the users (big data computing; services space for individuals, groups and organizations; self-learning systems; connecting people and society in the Internet of Things; etc.); (2) directions of interest for research (guaranteeing enhanced security, achieving expressivity of applications for future clouds, developing an efficient management in the computing environment, ensuring the reliability of the cloud systems, etc.); (3) possible changes in the infrastructure (multi-cloud, ad hoc cloud, heterogenous cloud, etc.); and (4) emergence of computing architectures (serverless computing, software-defined computing, etc.).[89]

(II) Looking the cloud computing from inside, there is another type of prediction, because the very *contents* that are stored in the cloud can be used for making predictions. They might be ontological, epistemological or heuristic predictions.[90] Thus, data analytics can be used for predicting phenomena, such as the outbreak of the dengue virus in Singapore, the behavior of a service or a user that is actually going on and has not been observed yet[91] or the trajectory of the new wave of the coronavirus pandemic. In this regard, it seems that future research directions on the cloud content include the following: 'The main goal is to transform the cloud from being a data management and infrastructure platform to scalable data analytics platform' (Vasuki et al. 2018, p. 3844). This is based on the idea of data as the new oil and 'the fundamental reason why cloud-based analytics are such a big thing is their easy accessibility, cost-effectiveness and ease of setting up and testing' (Vasuki et al. 2018, p. 3844).

Big data always appears in the future scenario for making predictions using contents available in the cloud, which includes prediction in the scientific, technological and social realms. There is data deluge in the cloud based on several kinds of sources: (i) operational data of organizations as result of business processes, monitoring the activities of the clients, web site tracking of their products, financial movements, etc.; (ii) the constant use of the social network Web sites by individuals and groups, which make explicit activities that they performed, events that they attend, things that they enjoy, prefer or want, etc.; and (iii) institutional information of different kinds, some available for the public, based on the data of the citizens. Big data is an expression that conveys the challenges that data deluge 'poses on existing infrastructure with respect the storage, management, interoperability, governance, and analysis of the data' (Assuncao et al. 2015, p. 4).

2.4.2 *The Practical Applications (Apps)*

Apps, or practical applications, are part of this third main layer. In their configuration, they 'rely on the same web architectures' as web browsing.[92] But there are apps that are connected to the Web, while others are not. According to Tim Berners-Lee, this structural and dynamic feature of the apps requires attention, because of its scientific, technological and social relevance for the present and the future:

> Some think that the future of media lies with phone apps, but a major issue with these is that the content is not on the Web. There is a huge difference with a Web app, which allows users to share URIs to get directly to underlying data sources, exploiting the added value of putting content on the Web. There is nothing wrong with phone apps per se, but none of the information in a phone app can be linked to. They are not participating in the new world of people linking things together, so for example, search engines will not find their content. There are guidelines for the creation of Web apps, to enable their function on all kinds of devices, from the Mobile Web Initiative at the W3C.[93]
>
> (Berners-Lee and O'Hara 2013, p. 2)

Berners-Lee with Kieron O'Hara 'sketch an infrastructure for Web apps, and consider issues pertaining to 'webizing' data. The basic plan for webizing data is to take the identifiers in a system and turn them into URIs; not all systems would survive such a change, but those that do will become much more powerful (Berners-Lee and O'Hara 2013, p. 2).[94] *De facto*, already in the year 2013, we had the following situation of massive use of apps through the mobile Internet:

> the success of smartphones has resulted in an explosive increase of apps available in app marketplaces. Currently, Apple's iOS app store and Google's play store both have more than 700,000 apps available while the Window Phone app store is catching up with more than 150,000 available apps. Among the most popular apps in the marketplace are real-time content-driven applications such as News, Email, Facebook, Twitter, and others that provide timely information to users.
>
> (Parate, Böhmer, Chu, Ganesan and Marlin 2013, p. 275)[95]

Considered philosophically, the issue of app design and its exponential growth through the mobile Internet highlights the existence of pragmatic complexity, which accompanies the structural complexity and dynamic complexity. This is an area where pragmatic complexity is emphasized, because it is about being able to deal with a wide variety of situations of a local, regional, national or international kind. All types of companies have been creating a wide variety of possibilities to deal with individual, group, organizational or institutional issues. This involves a type of complexity that adapts to the environment, which is also embedded by historicity, because the apps have to be updated as the practical issues address change.

Pragmatic complexity can also be available in cloud computing when dealing with big data and analytic solutions. In this regard, 'current solutions for analytics are often based on proprietary appliances or software systems built for general purposes. Thus, significant effort is needed to tailor such solutions to the specific needs of the organization, which includes integrating different data sources and deploying the software on the company's hardware (or, in the case of appliances, integrating the appliance hardware with the rest of the company's systems). Such solutions are usually deployed and hosted on the customer's premises, are generally complex, and their operations can take hours to execute' (Assuncao, Calheiros, Bianchi and Netto 2015, p. 4).

As for the future of the apps, in addition to those related to the Web and those directly linked to the mobile Internet, it is worth thinking about the increase of apps rooted in cloud platforms. This option would make it possible to customize or singularize services even more.[96] Some cloud platforms offer a 'notion of cloud apps and cloud apps marketplace (NetEx; Managing VMWare vApp – VMWare vSphere 4 ESX and VCenter Server; The Cloud Market; Complete Catalogue EC2 Images; Cisco – SourceFire). These 'apps' are primarily VMs [virtual machines] installed with pre-figured software stacks for a variety of standard work-flows, such as Web servers and databased servers. They primarily benefit clients who lack expertise or manpower to perform detailed configurations' (Nguyen et al. 2016, p. 178).

Then there is also the possibility of services as apps. This means 'a cloud market where apps (implemented by VMs) offer standard utilities such as firewalls, NIDS, storage encryption, and VMI [Virtual Machine Introspection]-based security tools' (Nguyen, Ganapathy, Srivastava and Vaidyanathan 2016, p. 178).[97]

2.4.3 *The 'Mobile Internet'*

Constitutively, the third major layer of the Internet – understood in a broad sense – has cloud computing and practical applications (apps) but also includes the 'mobile Internet.'[98] This area of *mobile Internet*, which has a growing development, is about mobile devices (mainly G4 and G5 smartphones) that are related to the network of networks. These 'cellular data networks' have had a great expansion in recent years. But there is some concern about the *quality* of the service. For both now and the future, the focus is on improving connectivity. In this regard, mobile devices do not have the adequate capacity to 'select the optimal network to enhance the quality of the service' (Yap et al. 2020, p. 1).

If this mobile setting is considered from a philosophico-methodological viewpoint, then the *heuristic prediction* comes into play. In this case, it has to be at least dual: (a) the prediction regarding the possible future in the scope of *connectivity* at a given time of the mobile device at stake; and (b) the prediction concerning the expected *behavior* of the users of this device (individuals, groups or organizations), which may need to know if there is connectivity and the degree of reliability of access to the data (personal, professional or institutional) in the network of networks.

Complexity also comes into play, because connectivity can be to two different types of wireless architectures: (i) homogeneous networks, which facilitate reliability of connectivity and predictability in access to information and (ii) heterogeneous networks, which decrease that reliability in the sought-after connectivity to the network of networks. All these elements should contribute to a better understanding of how the network service provide should perform the task, in order to guarantee the quality of access and avoid its degradation for users.[99] Even so, this complexity will continue to increase, generating problems not previously contemplated in this third layer of the Internet in the broad sense:

The growth in mobile and wireless networks has disrupted the way that networks were designed. A recent report showed that mobile data usage grew 63% in 2016 and that it will surpass 49 exabytes by 2021.[100] Smart phones have eclipsed personal computers, and the global population of over 7.4 billion wireless users continues to consume an increasing amount of spectrum resources. Cloud-based services and video traffic as well as essential online services such as e-banking, e-learning and e-health continue to proliferate, causing the wireless traffic volume to grow exponentially.

(Yap et al. 2020, p. 1)

Faced with the problem of achieving better connectivity, two aspects – internal and external – are combined in this area. (a) The *internal* perspective of improving mobile devices is mobilized by the demand for better performance. In this regard, it happens that, 'since mobile devices cannot predict and optimize the handover process, the quality of services (QoS) for mobile users degrades further, especially when mobile users are on the move' (Yap et al. 2020, p. 2). (b) The *external* perspective of the social dimension is when the users demand access to the network 'anytime and anywhere' as well as a better quality in the service provided.

De facto, there are at least *two kinds of predictions* involved here. On the one hand, there is an *artificial* type of prediction, dealing with processes, which has to determine the scope and reliability for potential connectivity by the available devices. These are, in principle, predictions of a heuristic rather than ontological or epistemological nature.[101] On the other hand, there is a *social* type of prediction, which has to anticipate the possible future behavior of the users of the devices (individuals, groups or organizations) based on the knowledge of the past trajectory of their behavior.

Furthermore, there are two possible routes for dealing with this problem from prediction: in *homogeneous* networks and in *heterogeneous* networks. In the first case, mobility prediction – based on knowledge of user behavior – can reliably anticipate the location of these agents, improve the handover process and manage resources efficiently. Meanwhile, in the second case, a reliable outcome is more difficult. In heterogeneous networks, a multipath scheme can be used in order to select the quality of service by means of the multipath transmission control protocol. This protocol discovers the number of available paths that can be at hand for the users and 'establishes the paths and distributes traffic across these paths through the creation of separate sub-flows based on the lowest RTT [round trip time] and unfilled congestion window' (Yap et al. 2020, p. 2). However, this may not be a satisfactory solution in some cases.

2.5   Coda: From the Present Situation to the Near Future

Nowadays, we have *the* Internet, which is a version of *an* internet, as D. Clark insists,[102] and this chapter considers its possible future. This present network of networks, which has had a worldwide scientific, technological and social impact through its three main layers, has brought with it a new historical stage. This can be called 'Hyperhistory'[103] or can be given another name, such as 'Knowledge Society' – that fits with the social dimension of the network of networks – or even 'Information Age,' which seems to me a less specific denomination in terms of space and time. De facto, this historical component of the social dimension of the Internet – in the broad sense – has been highlighted in the confinement due to the Covid-19 pandemic.[104] Thus, at least for countries affected by confinement, there is a before and after use of the network of networks for social life in all its forms: health, commerce, education, entertainment, etc.

But the very success of the network of networks, especially since it became accessible to everyone and became the largest commercial support on the planet,[105] creates internal and external problems today and may lead to more in the near future. These problems are more visible when they affect the social dimension, which has a political side, such as dissemination of misinformation (mainly through social networks, such as Facebook or Twitter), the spread of disinformation[106] or the rights on patents due to geopolitical issues (as has happened with G5 technology and the mobile Internet, because of the patents of the Chinese company Huawei).

Concerning those current problems that may have future projection, we should start from the idea of the complexity of the phenomenon, which includes structural, dynamic and pragmatic complexity. This complexity of the system is what this chapter has sought to highlight in its analysis. It is a complexity noticeable in the interaction of the internal and external levels of the network of networks, which arises from the interdependence between the scientific side, the technological side and the social dimension. Each of them influences the epistemological terrains of the short, medium and long term but also the ontological macro, meso and micro levels of this complex system.

Taken as a whole, the present situation and the possible future of the Internet in the broad sense depend on the interdependence between the scientific side, the technological side and the social dimension levels, which occur at the three main layers studied here: the technological network (the backbone and service provider), the Web (with social networks and an increasing number of possibilities) and the cloud computing, apps and the mobile Internet. They should be investigated from a methodological pluralism in order to face these new problems with novel approaches that should be effective for the solution of the challenges arising from science, technology and society.

Notes

1 This paper has been written within the framework of research project PID2020-119170RB-I00 supported by the Spanish Ministry of Science and Innovation (AEI). Previously, some of the ideas in this text were presented at the *Grappling with the Futures Symposium*, organized by Harvard University and Boston University, and at the conference *For a Bottom-Up Epistemology*, organized by the University of Bologna.

2 Cf. Clark (2018), p. 37, Schultze and Whitt (2016), and The Economist (2018).

3 'The Internet should not be confused with the World Wide Web, a virtual network of webpages and hyperlinks' (Newman 2018, p. 15). An overview of the current situation and possible future of the Web can be found in Berendt et al. (2020).

4 In this interaction, the scientific side intervenes in terms of applied science and the application of science, which requires prediction and prescription, cf. Gonzalez (2018a).

5 This service to society has been evident since the beginning of the Covid-19 pandemic, cf. Gonzalez (2020b). However, there are relevant problems, among which there are two types that are particularly important for the social dimension of the Internet in a broad sense. Those are the ones related to cybersecurity and to fake news (disinformation and misinformation).

6 Dynamic complexity begins on the Internet as a platform for information and communication. Cf. Gonzalez (2018b).

7 One issue is to propose solutions to specific problems, as applied science does, and another is to specify those solutions in contexts of use that are variable and have unique characteristics, which is what application of science does. Cf. Gonzalez (2020c), pp. 251–257, 259, 262–267, 270, 273, 275–276, and 279n–280.

8 These modes of complexity with general character appear in Rescher (1998), pp. 1–24; especially, p. 9. For the case of complexity in the Internet in a broad sense, cf. Gonzalez (2022).

9 It belongs to the field of the sciences of the artificial insofar as it is goal-oriented and seeks the empowerment of the human agents. The starting point of the design sciences is thematized in Simon (1996).

10 'Internet Protocol (IP) sits in the "middle" of all Internet communications. Any device or application that connects to the Internet ultimately have its communications translated into this lingua franca of the network. However, IP is highly extensible because it allows for new virtual protocols to be built "above" this layer, or for new physical devices to implement ways to transport IP "below" its layer. Higher-layer protocols include HTTP (a standard for web traffic), and lower-layer protocols include Ethernet (a standard for wired devices). This is what makes up the so-called "hourglass" structure of the Internet' (Schultze and Whitt 2016, p. 64).

11 Cf. Henten and Tadayoni (2008), p. 52.

12 Cf. Henten and Tadayoni (2008), p. 53.

13 Cf. Henten and Tadayoni (2008), p. 54.

14   'A module has some specified *interface* through which it connects to the rest of the system, and the internals of the module, beneath the interface, are hidden and not accessible from outside the module' (Clark 2018, p. 35).

15   Cf. Schultze and Whitt (2016), p. 65.

16   Cf. Meeker (2019).

17   Cf. Clark (2018), pp. 145 and 148. They are also known as '*network backbone providers* (NBP), who are primarily national governments and major telecommunication companies such as Level 3 Communications, Cogent, NTT, and others' (Newman 2018, p. 15).

18   Cf. Clark (2018), pp. 6, 102, and 242–245.

19   The second can be 'subdivided into *regional ISPs* and *local* or *consumer ISPs*, the former being larger organizations whose primary customers are the local ISPs, who in turn sell network connections to the end users. This distinction is somehow blurred however, because larger consumer ISPs, such as AT&T or British Telecom, often act as their own regional ISPs (and some may be backbone providers as well)' (Newman 2018, p. 16).

20   Cf. Newman (2018), pp. 15–16. In addition to the structural component, there is the dynamic one. In this regard, 'link prediction can provide a useful methodology for the modeling of networks. The evolving mechanisms of networks have been widely studied. Many evolving models have been proposed to capture the evolving process of real-world networks. However, it is very hard to quantify the degree to which the proposed evolving models govern real networks. Actually, each evolving model can be viewed as the corresponding predictor, we can thus apply evaluating metrics on prediction accuracy to measure the performance of different models' (Tan et al. 2014, p. 1).

21   The governance of the Internet, both in the strict sense (the infrastructure) and in the broad sense (the network of networks), can be approached from the internal point of view (the supervision of the very functioning of the infrastructure, new rules for its development and adaptation to novel circumstances, etc.) or from an external perspective, starting from the public environment and reaching international relations. The first line is highlighted by books such as Bygrave and Bing (2009). Meanwhile, the second line predominates in volumes such as Goldsmith and Wu (2006), Mueller (2013), and DeNardis (2014).

22   'The birth of the Internet is (…) a worldwide network arose from the sequential adding of links among nodes (computers)' (Bagnoli et al. 2019, p. 2).

23   On these scientific disciplines, see Tiropanis et al. (2015).

24   Cf. Gonzalez (2018a).

25   Cf. Gonzalez (2015b).

26   From a historical point of view, the uniqueness of the Internet as a network should be highlighted. See, in this respect, how it is described by the main protagonists: Leiner et al. (1997) and Leiner et al. (2009).

27   Cf. Tiropanis et al. (2015), pp. 76–82; especially, pp. 76–79 and 81–82.

28   'Russia wants to create a "sovereign internet" that can be cut from the rest of the online world at the flip of a switch (…). In a splinternet world choice will be limited, costs will rise and innovation will slow. And all the while China, with the largest access to data, loses least' (The Economist 2020d, p. 19). On the matter of digital infrastructures and data sovereignty, see EIT Digital (2020).

29   Cf. Faloutsos et al. (1999), and Holme et al. (2004).

30   Cf. Ebel et al. (2002), and Holme et al. (2004).

31   'The resilience of the Internet to failure (…) implies the presence of heavily connected hubs' (Bagnoli et al. 2019, p. 2).

32 Cf. Albert et al. (2000).
33 Because the Internet of Things has a public and operational dimension, it involves new legal problems that are now under discussion in conferences, cf. Zhang and Zhang (2017).
34 A key issue of the IoT is interconnectivity, cf. Ornes (2016).
35 On the differences between human intelligence and artificial intelligence, see Gonzalez (2017).
36 Alphabet (Google) is, at the same time, a world leader in Artificial Intelligence and self-driving cars. See The Economist (2020a), p. 8.
37 Cf. Palmieri (2013).
38 It was that day when 'the European Organization for Nuclear Research (CERN) announced that the World Wide Web it had created would be free to everyone' (Floridi 2014, p. 18).
39 Cf. Gonzalez (2020a).
40 Cf. Robertson and Giunchiglia (2013).
41 Cf. Shadbolt et al. (2013), p. 3.
42 Cf. Kermarrec (2013).
43 This trend towards customization in the Web layer contributes to the presence of pragmatic complexity, as it implies a clearly bounded singularization or specification.
44 That the development of the Web has been an open process rather than clearly prefixed goals is proven by the successive versions of the Web, now known in a synthetic way as Web 1.0, Web 2.0, Web 3.0, etc.
45 'The Web is a social phenomenon whose vast scale has produced emergent properties and transformative behaviours. The Web is a space built and used by people for people. We will not understand it if we simply reduce it to its technological parts' (Shadbolt et al. 2013, p. 1).
46 On the social dimension of the Internet, see Gonzalez (2020a).
47 'Understanding the Web, its current and future trajectories through AI, demands that we craft new approaches at the intersections of traditional disciplinary practice. As the object of our study evolves and the necessity for collaboration across disciplines and with researchers in industry and government becomes ever-greater, it is urgent to that we innovate our methods and research infrastructures accordingly' (Berendt et al. 2020, p. 24).
48 Cf. Shadbolt et al. (2020), p. 1.
49 On the Internet and the social sciences, see Ackland (2013), and Askitas and Zimmermann (2015).
50 Cf. Berners-Lee et al. (2006) and Hendler and Hall (2016).
51 'Scientification' is used here in the sense of Ilkka Niiniluoto. See Niiniluoto (1993).
52 This is a science of the artificial in the research line opened up by the philosophico-methodological approach initiated by Herbert Simon in his book *The Sciences of the Artificial*, initially published in 1969, which had a second edition in 1981 and was expanded in the third edition in 1996.
53 Cf. Gonzalez (2020b).
54 On the sciences of the Internet, such as the Web science, as sciences of the artificial, see Gonzalez (2018a, 2018b) and Gonzalez and Arrojo (2019).
55 Some of the founders of the Web science insist on the 'interdisciplinary drawing insights from mathematics, physics, computer science, psychology, ecology, sociology but also law, political science, economics and more' (Shadbolt et al. (2013), p. 5).
56 Cf. Shadbolt et al. (2013), p. 5.
57 In 2018 and 2019 it has been on display at the Science Museum in London.
58 On the issue of the emerge in the Web, see O'Hara et al. (2012).

59   For Robert M. May, the Web as a whole may not be 'scale-free.' It cannot simply be assumed that a sample of this network represents the degree of distribution of the Web as such. When there is an attack, the response depends not only on the topology but also on the degree of strength or weakness of the individual links that are being disrupted. Cf. May (2013).

60   Cf. Barabási (2013), pp. 1–3.

61   'The *small world* property, which says that two nodes are likely to be connected, even in such a very large and sparse scale-free network as the Web, by a relatively short path of nodes – in the case of the Web, the path length is about 19' (Barabási 2013, p. 2).

62   Cf. Berendt et al. (2020), pp. 1 and 30.

63   One way forward to reduce the negative and promote sustainability comes from the self-regulation of enterprises. 'Examples include Tumblr forbidding pornographic content. Instagram limiting the viewing of number of likes and Twitter denying political advertisements' (Berendt et al. 2020, p. 34, note 20).

64   Cf. von Ahn (2013).

65   At the same time, 'AI tools also hold various potential for Web science research itself, enable us to observe, analyse and intervene in the evolution of the Web and may even serve as boundary objects for interdisciplinary processes' (Berendt et al. 2020, p. 26).

66   Cf. von Ahn (2013), p. 3.

67   Following this line, the problem of fake news will increase the need for individual and social responsibility.

68   'When it comes to digitization, health care has indeed lagged behind not just banking but travel, retail, carmaking and even packed goods. Some 70% of America hospitals still fax and post patient records. (…). Patients are growing more comfortable with remote and computer-assisted diagnosis and treatment. And enterprising firms, from health-app startups and hospitals to insurers, pharmacies and tech giants such as Amazon, Apple and Google, are scrambling to provide such services' (The Economist 2020e, p. 54).

69   'Under lockdown, e-commerce as a share of America retail sales increased as much in eight weeks as it had in the previous five years' (The Economist 2020f, p. 15).

70   This issue affects not only the Web layer but also the other two layers of the Internet in a broad sense: 'There are many unsolved security threats on today's Internet, such as malware, phishing of (bank) account data, and denial of service attacks. (…) In the future Internet the situation will become more complicated. Web-based applications will require access to the user's data, and Web browsers will partially replace operating systems; therefore, security vulnerabilities in browsers or browsers plugins could expose the user's locally and remotely stored data to attacks' (Schönwälder et al. 2009, p. 31).

71   'It is necessary to stress that traditional security approaches, which are more or less centralized fail in such distributed environments, and that new approaches in terms of ambient security are needed. The aspect of the quality of security needs to be approached as well' (Schönwälder et al. 2009, p. 30).

72   The issue of the copyright is particularly relevant in some contexts related to the Web. See Freedman et al. (2008).

73   Cf. Hendler (2023), pp. 78–82.

74   Cf. GDPR (2016).

75   Cf. Rokhanna (2020).

76   Cf. Hendler (2023), p. 82.

77   Cf. Simon (1997), pp. 295–298.

78   Cf. Rescher (1987).

79 'Azure aims to match or overtake AWS in the cloud. (...) The way Microsoft has built its global cloud infrastructure, covering more geographical ground than AWS but more thinly, may it less reliable. (...) As demand has surged in the pandemic, with millions of remote workers switching to the cloud, Azure was at times been unable to keep up. Microsoft Teams suffered a blackout in March [2020]' (The Economist 2020c, p. 61).

80 'The cloud provides a good platform for big data storage, processing and analysis, addressing two main requirements of big data analytics, high storage and high performance computing' (Vasuki et al. 2018, p. 3842).

81 'Infrastructure-as-a-service is a model that provides computing and storage resources as a service. On the hand, in case of PaaS and SaaS, the cloud services provide software platform or software itself as a service to its clients' (Vasuki et al. 2018, p. 3842).

82 Cf. Azure (2020).

83 Cf. Clark (2018), p. 307.

84 Cf. Vasuki et al. (2018), p. 3843.

85 Regarding the future of the cloud computing and the big data (in terms of variety, velocity, volume, veracity and value), see Mohapatra et al. (2015).

86 There are, in fact, already companies that can use data from several of the above-mentioned clouds.

87 'The services of the future Internet require us to think of management issues already at the design phase of new services' (Schönwälder et al. 2009, p. 29).

88 'The future Internet and the provided future services will change our lives, and especially the way we communicate and use our environment. Personalized services, data, and content will be accessible from everywhere by using heterogenous equipment adapting automatically to the specifics of the underlying networks and the surrounding infrastructure' (Schönwälder et al. 2009, p. 29).

89 Cf. Varghese and Buyya (2018), pp. 849–861; especially, p. 850.

90 On these three kinds of predictions, see Gonzalez (2015a), pp. 108–112.

91 Cf. Vasuki et al. (2018), p. 3843.

92 Cf. Hendler and Hall (2016), p. 704.

93 See also W3C Mobile Initiative (2021).

94 Berners-Lee and O'Hara exemplify their view in figure 2: 'apps – Web apps, trusted apps on a computer, public apps or apps using a person's credentials on his or her machine – sit on top. With systems such as these, with their assumptions about social context as well as technology, it is clearly important to control access. Hence, this is properly thought of as *read–write access-controlled linked data*, or, put another way, *socially aware cloud storage*. The structure is analogous to UNIX storage burst open onto the Web, functioning at Web scale' (Berners-Lee and O'Hara 2013, p. 4).

95 The paper was presented on 8 September 2013.

96 Cf. Nguyen et al. (2016).

97 In this regard, 'using VMI one can build a rootkit detectors for operating systems executing within the VM [Virtual Machine]' (Nguyen et al. 2016, p. 178).

98 Cf. Yap et al. (2020), p. 1.

99 Cf. Yap et al. (2020), p. 1.

100 Cf. Cisco (2019).

101 On the distinction between heuristic, ontological and epistemological predictions, see Gonzalez (2015a).

102 'This is a book about how to design an internet. I say *an internet* rather than *the Internet* because the book is about not just the Internet we have today

but also possible alternative conceptions of an internet – what we might have designed back then or might contemplate in the future' (Clark 2018, p. 1).

103 Cf. Floridi (2014), pp. 1–24.

104 Cf. Gonzalez (2020b), pp. 383–412; especially, p. 393.

105 On how the network of networks became the great commercial support, see Greenstein (2015).

106 'Mis- and disinformation existed before the Web, but the Web has facilitated its distribution, at scale and speed. Artificial Intelligence further amplifies the problem. (…) Artificial Intelligence allows the creation of *deep fakes*, e.g. video or audio faking the voice and the appearance of anybody while being indistinguishable from true video or audio' (Berendt et al. 2020, p. 8). But Artificial Intelligence can also contribute to tasks such as fact checking, fraud detection and protection enforcement to support privacy, cf. Berendt et al. (2020), p. 14.

References

Ackland, R. 2013. *Web Social Science: Concepts, Data and Tools for Social Scientists in the Digital Age*, London: SAGE.

Albert, R., Jeong, H. and Barabási, A.-L. 2000. 'Error and Attack Tolerance of Complex Networks', in *Nature*, vol. 406/6794, pp. 378–382. DOI: 10.1038/35019019.

Askitas, N. and Zimmermann, K. F. 2015. 'The Internet as a Data Source for Advancement in the Social Sciences', in *International Journal of Manpower*, vol. 36/1, pp. 2–12.

Assuncao, M. D., Calheiros, R. N., Bianchi, S. and Netto, M. A. S. 2015. 'Big Data Computing and Clouds: Trends and Future Directions', in *Journal of Parallel and Distributed Computation*, vol. 79–80, pp. 3–15. DOI: 10.1016/j.jpdc.2014.08.003.

Azure. 2020. 'What Is Cloud Computing?' Available at https://azure.microsoft.com/en-us/overview/what-is-cloud-computing/ Accessed at 24 August 2020.

Bagnoli, F., Bellini, E., Massaro, E. and Rechtman, R. 2019. 'Percolation and Internet Science', in *Future Internet*, vol. 11/35, pp. 1–26. DOI: 10.3390/fi11020035.

Barabási, A. L. 2013. 'Network Science', in *Philosophical Transactions of the Royal Society A*, vol. 371/1987, pp. 1–3. DOI: 10.1098/rsta.2012.0375.

Berendt, B., Gandon, F., Halford, S., Hall, W., Hendler, J., Kinder-Kurlanda, K. E., Ntoutsi, E. and Staab, S. 2020. 'Web Futures: Inclusive, Intelligent, Sustainable. The 2020 Manifesto for Web Science', in *Manifesto from Dagstuhl Perspectives Workshop 18262*, pp. 1–42. Available at: https://drops.dagstuhl.de/opus/volltexte/2021/13744/ Accessed at 4 May 2021.

Berners-Lee, T., Hall, W., Hendler, J., Shadbot, N. and Weitzner, D. J. 2006. 'Creating a Science of the Web', in *Science*, vol. 313/5788, pp. 769–771.

Berners-Lee, T. and O'Hara, K. 2013. 'The Read-write Linked Data Web', in *Philosophical Transactions of the Royal Society A*, vol. 371/1987, pp. 1–5. DOI: 10.1098/rsta.2012.0386.

Bygrave, L. A. and Bing, J. (eds.). 2009. *Internet Governance: Infrastructure and Institutions*, Oxford: Oxford University Press.

Clark, D. D. 2018. *Designing an Internet*, Cambridge, MA: The MIT Press.

Cisco. 2019. *Cisco Visual Networking Index: Global Mobile Data Traffic Forecast Update*, 2017–2022. White Paper, February, 33 pages. Available at: https://s3.amazonaws.com/media.mediapost.com/uploads/CiscoForecast.pdf Accessed at 19 August 2020.

DeNardis, L. 2014. *The Global War for Internet Governance*, New Haven, CT: Yale University Press.

Ebel, H., Mielsch, L. and Bornholdt, S. 2002. 'Scale-Free Topology of E-Mail Networks', in *Physical Review, E. Statistical, Nonlinear, and Soft Matter Physics*, vol. 66, 30 September, 035103. DOI: 10.1103/PhysRev.66.035103

EIT Digital. 2020. *European Digital Infrastructure and Data Sovereignty: A Policy Perspective*, EIT, European Union, Brussels, 9.6.2020. Available at https://www.eitdigital.eu/fileadmin/files/2020/publications/data-sovereignty/EIT-Digital-Data-Sovereignty-Summary-Report.pdf Accessed at 12 February 2021.

Faloutsos, M., Faloutsos, P. and Faloutsos, C. 1999. 'On Power-Law Relationships of the Internet Topology', in *Proceedings of the Conference on Applications, Technologies, Architectures, and Protocols for Computer Communication*, Cambridge, MA, 30 August–3 September 1999. Association for Computing Machinery, New York, NY. (See *Computer Communication Review*. 1999. v. 29, n. 4, pp. 251–262.)

Floridi, L. 2014. *The Fourth Revolution – How the Infosphere is Reshaping Human Reality*, Oxford: Oxford University Press.

Freedman, D., Henten, A., Towse, R. and Wallis, R. 2008. 'The Impact of the Internet on Media Policy, Regulation and Copyright Law', in L. Küng, R. G. Picard and R. Towse (eds.), *The Internet and the Mass Media*, London: SAGE, pp. 102–124.

GDPR 2016. 'General Data Protection Regulation', in *Official Journal of the European Union*, vol. 59, 4 May, pp. 1–88. Available at: https://gdpr-info.eu Accessed at 19 November 2019.

Goldsmith, J. and Wu, T. 2006. *Who Controls the Internet? Illusions of a Borderless World*, Oxford: Oxford University Press.

Gonzalez, W. J. 2015a. *Philosophico-Methodological Analysis of Prediction and its Role in Economics*, Dordrecht: Springer.

Gonzalez, W. J. 2015b. 'On the Role of Values in the Configuration of Technology: From Axiology to Ethics', in W. J. Gonzalez (ed.), *New Perspectives on Technology, Values, and Ethics: Theoretical and Practical*, Boston Studies in the Philosophy and History of Science, Dordrecht: Springer, pp. 3–27.

Gonzalez, W. J. 2017. 'From Intelligence to Rationality of Minds and Machines in Contemporary Society: The Sciences of Design and the Role of Information', in *Minds and Machines*, vol. 27/3, pp. 397–424. DOI: 10.1007/s11023-017-9439-0. Available at: https://link.springer.com/article/10.1007/s11023-017-9439-0 Accessed at 6 October 2017.

Gonzalez, W. J. 2018a. 'Internet en su vertiente científica: Predicción y prescripción ante la complejidad', in *Artefactos: Revista de Estudios sobre Ciencia y Tecnología*, vol. 7/2, 2nd period, pp. 75–97. DOI: 10.14201/art2018717597.

Gonzalez, W. J. 2018b. 'Complejidad dinámica en Internet como plataforma de información y comunicación: Análisis filosófico desde la perspectiva de Ciencias de Diseño y el papel de la predicción', in *Informação e Sociedade: Estudos*, vol. 28/1, pp. 155–168.

Gonzalez, W. J. and Arrojo, M. J. 2019. 'Complexity in the Sciences of the Internet and Its Relation to Communication Sciences', in *Empedocles: European Journal for the Philosophy of Communication*, vol. 10/1, pp. 15–33. DOI: 10.1386/ejpc.10.1.15_1 Available at: https://www.ingentaconnect.com/contentone/intellect/ejpc/2019/00000010/00000001/art00003 Accessed at 6 July 2019.

Gonzalez, W. J. 2020a. 'La dimensión social de Internet: Análisis filosófico-metodológico desde la complejidad', in *Artefactos: Revista de Estudios de la Ciencia y la Tecnología*, vol. 9/1, 2nd period, pp. 101–129. DOI: 10.14201/art2020101129. Available at: https://revistas.usal.es/index.php/artefactos/article/view/art2020101129 Accessed at 27 April 2020.

Gonzalez, W. J. 2020b. 'The Internet at the Service of Society: Business Ethics, Rationality, and Responsibility', in *Éndoxa*, vol. 46, pp. 383–412. Available at: http://revistas.uned.es/index.php/endoxa/article/view/28029/pdf

Gonzalez, W. J. 2020c. 'Pragmatic Realism and Scientific Prediction: The Role of Complexity', in W. J. Gonzalez (ed.), *New Approaches to Scientific Realism*, Boston/Berlin: De Gruyter, pp. 251–287. DOI: 10.1515/9783110664737-012.

Gonzalez, W. J. 2022. 'Scientific Side of the Future of the Internet as a Complex System. The Role Prediction and Prescription of Applied Sciences', in W. J. Gonzalez (ed.), *Current Trends in Philosophy of Science. A Prospective for the Near Future*, Synthese Library. Cham: Springer, pp. 103–144.

Greenstein, S. 2015. *How the Internet Became Commercial. Innovation, Privatization, and the Birth of a New Network*, Princeton, NJ: Princeton University Press.

Hendler, J. and Hall, W. 2016. 'Science of the World Wide Web', in *Science*, vol. 354/6313, pp. 703–704.

Hendler, J. 2023. 'The Future of the Web', in W. J. Gonzalez (ed.), *The Internet and Philosophy of Science*. Routledge Studies in the Philosophy of Science, London: Routledge, pp. 71–83.

Henten, A. and Tadayoni, R. 2008. 'The Impact of the Internet on Media Technology, Platforms and Innovation', in L. Küng, R. G. Picard and R. Towse (eds.), *The Internet and the Mass Media*, London: SAGE, pp. 45–64.

Holme, P., Edling, C. R. and Liljeros, F. 2004. 'Structure and Time Evolution of an Internet Dating Community', *Social Network*, vol. 26/2, pp. 155–174. DOI: 10.1016/j.socnet.2004.01.007.

Kermarrec, A. M. 2013. 'Towards a Personalized Internet: A Case for Full Decentralization', in *Philosophical Transactions of the Royal Society A*, vol. 371/1987, pp. 1–17. DOI: 10.1098/rsta.2012.0380.

Leiner, B. M., Cerf, V. G., Clark, D. D., Kahn, R. E., Kleinrock, L., Lynch, D. C., Postel, J., Roberts, L. G. and Wolff, S. 1997. 'The Past and Future History of the Internet. The Science of the Future Technology', in *Communications of the ACM*, vol. 40/2. pp. 102–108.

Leiner, B. M., Cerf, V. G., Clark, D. D., Kahn, R. E., Kleinrock, L., Lynch, D. C., Postel, J., Roberts, L. G. and Wolff, S. 2009. 'A Brief History of the Internet', in *ACM SIGCOMM Computer Communication Review*, vol. 39/5, pp. 22–31.

May, R. M. 2013. 'Networks and Webs in Ecosystems and Financial Systems', in *Philosophical Transactions of the Royal Society A*, vol. 371/1987. DOI: 10.1098/rsta.2012.0376.

Meeker, M. 2019. *Internet Trends 2019*. Report published on June 11, 334 pages. Available at: https://www.bondcap.com/pdf/Internet_Trends_2019.pdf Accessed at 22 July 2019.

Mohapatra, S., Paikaray, J. and Samal, N. 2015. 'Future Trends in Cloud Computing and Big Data', in *Journal of Computer Sciences and Applications*, vol. 3/6, pp. 137–142. DOI: 10.12691/jcsa-3-6-6.

Mueller, M. L. 2013. *Networks and States. The Global Politics of Internet Governance*, Cambridge: The MIT Press.

Newman, M. 2018. *Networks*, Oxford: Oxford University Press, 2nd ed.

Nguyen, H., Ganapathy, V., Srivastava, A. and Vaidyanathan, S. 2016. 'Exploring Infrastructure Support of App-based Services on Cloud Computing', in *Computers and Security*, vol. 62/4, pp. 177–192.

Niiniluoto, I. 1993. 'The Aim and Structure of Applied Research', in *Erkenntnis*, vol. 38/1, pp. 1–21.

O'Hara, K., Contractor, N. S., Hall, W., Hendler, J. A. and Shadbolt, N. 2012. 'Web Science: Understanding the Emergence of Macro-Level Features on the World Wide Web', in *Foundations and Trends in Web Science*, vol. 4/2-3, pp. 103–267. DOI: 10.1561/1800000017.

Ornes, S. 2016. 'The Internet of Things and the Explosion of Interconnectivity', in *Proceedings of the Natural Academy of Sciences* (USA), vol. 116/4, pp. 11.059–11.060.

Palmieri, F. 2013. 'Scalable Service Discovery in Ubiquitous and Pervasive Computing Architectures: A Percolation-driven Approach', in *Future Generation Computer Systems*, vol. 29/3, pp. 693–703.

Parate, A., Böhmer, M., Chu, D., Ganesan, D. and Marlin, B. 2013. 'Practical Prediction and Prefetch for Faster Access to Applications on Mobile Phones', in *Proceedings of the 2013 ACM International Joint Conference on Pervasive and Ubiquitous Computing*, New York: Association for Computing Machinery, pp. 275–284. DOI: 10.1145/2493432.2493490.

Park, J. H. 2019. 'Advances in Future Internet and the Industrial Internet of Things', in *Symmetry*, vol. 11/2, pp. 1–4. DOI: 10.3390/sym11020244.

Rescher, N. 1987. *Ethical Idealism: An Inquiry into the Nature and Function of Ideals*, Berkeley: University of California Press.

Rescher, N. 1998. *Complexity: A Philosophical Overview*, New Brunswick, NJ: Transaction Publishers.

Robertson, D. and Giunchiglia, F. 2013. 'Programming the Social Computing', in *Philosophical Transactions of the Royal Society A*, vol. 371/1987, pp. 1–14. DOI: 10.1098/rsta.2012.0379.

Rokhanna 2020. 'Internet Bill of Rights'. Available at: https://www.rokhanna.com/issues/internet-bill-rights Accessed at 9 January 2020.

Schönwälder, J., Fouquet, M., Rodosek, G. D. and Hochstatter, C. 2009. 'Future Internet = Content + Services + Management', in *IEEE Communications*, vol. 47, pp. 27–33.

Schultze, S. J. and Whitt, R. S. 2016. 'Internet as a Complex Layered System', in J. M. Bauer and M. Latzer (eds.), *Handbook on the Economics of the Internet*, Cheltenham: Edward Elgar, pp. 55–71.

Shadbolt, N., Hall, W., Hendler, J. A. and Dutton, W. H. 2013. 'Web Science: A New Frontier', in *Philosophical Transactions of the Royal Society A*, vol. 371/1987, pp. 1–6 DOI: 10.1098/rsta.2012.0512.

Simon, H. A. 1996. *The Sciences of the Artificial*, Cambridge, MA: The MIT Press, 3rd ed. (1st ed. in 1969, and 2nd ed. in 1981).

Simon, H. A. 1997. 'Satisficing', in H. A. Simon (ed.), *Models of Bounded Rationality*. Vol. 3: *Empirically Grounded Economic Reason*, Cambridge, MA: The MIT Press, MA, pp. 295–298.

Tan, F., Xia, Y. and Zhu, B. 2014. 'Link Prediction in Complex Networks: A Mutual Information Perspective', in *PLoS ONE*, vol. 9/9, pp. 1–8. DOI: 10.1371/journal.pone.0107056.

The Economist 2018. 'More Knock-on Than Network. How the Internet Lost Its Decentralised Innocence', *Special Report: Fixing the Internet in The Economist*, vol. 427, n. 9098, June 30, pp. 5–6.

The Economist 2020a. 'Google: How to Cope with Middle Age', Section *Leaders*, August 1, p. 8.

The Economist 2020b. 'Some Lessons from Microsoft. Blue-sky Thinking', Section *Leaders*, October 24, p. 13.

The Economist 2020c. 'After the Reboot. The Software Giant Has Turned Itself Around. Now for the Hard Part', Section *Briefing: Microsoft*, October 24, pp. 60–62.

The Economist 2020d. 'The New Grand Bargain', Section *Briefing: Global technopolitcs*, November 21, pp. 17–20.

The Economist 2020e. 'Health Care and Technology. The Dawn of Digital Medicine', Section *Business*, December 5, pp. 54–55.

The Economist 2020f. 'The Plague Year. This Will Be Remembered as a Moment When Everything Changed', Section *Leaders*, December 19, p. 15.

Tiropanis, T., Hall, W., Crowcroft, J., Contractor, N. and Tassiulas, L. 2015. 'Network Science, Web Science, and Internet Science', in *Communications of ACM*, vol. 58/8, pp. 76–82.

Varghese, B. and Buyya, R. 2018. 'Next Generation Cloud Computing: New Trends and Research Directions', in *Future Generation of Computer Systems*, vol. 79, pp. 849–861. DOI: 10.1016/j.future.2017.09.020.

Vasuki, K., Rajeswari, K. and Prabakaran, M. 2018. 'A Survey of Current Research and Future Directions Using Cloud-Based Big Data Analytics', in *International Research Journal of Engineering and Technology*, vol. 5/5, pp. 3841–3844.

von Ahn, L. 2013. 'Augmented Intelligence: The Web and Human Intelligence', in *Philosophical Transactions of the Royal Society A*, vol. 371/1987, pp. 1–3. DOI: 10.1098/rsta.2012.0383.

World Wide Web Foundation. 2019. *Contract for the Web. A Global Plan of Action to Make our Online World Safe and Empowering for Everyone*, November 2019. Available at: https://contractfortheweb.org Accessed at 25 November 2019.

W3C Mobile Initiative. 2021. Available at http://www.w3.org/Mobile/ Accessed at 16 January 2021.

Yap, K. L., Chong, Y. W. and Liu, W. 2020. 'Enhance Handover Mechanism Using Mobile Prediction in Wireless Networks', in *PLoS ONE*, vol. 15/1, pp. 1–31. DOI: 10.371/journal.phone.0227982.

Zhang, J. and Zhang, W. 2017. 'Future of Law Conference: The Internet of Things, Smart Contracts and Intelligent Machines', in *Frontiers of Law in China*, vol. 12/4, pp. 673–674.

3 The Future of the Web

James A. Hendler

3.1 Introduction

Some thirty years have passed since one of the greatest societal changes in human history began. It took place when the inventor of the World Wide Web, Tim Berners-Lee, defined its technical foundations. Since then, the Web has grown exponentially. It has gone far beyond what was envisioned in its original technical foundations. It has also had a profound impact on the world today and will certainly have a greater impact on the society of the future. In a 2018 report, the UN Broadband Commission estimated that over half of the Earth's population would be connected to the Web by the end of 2019,[1] and recent numbers validate that estimate, putting the 2021 usage at close to 60% of the world's population.[2]

On the one hand, we have seen that the Web can influence the models pursued by business, the realization of human rights and even the pursuit of happiness. The Web can provide an infrastructure for multiple human and social tasks, such as those related to learning, working, communicating with loved ones, and having entertainment. On the other hand, the Web also creates an environment affected by the digital divide between those who have access and those who do not have access. And, not surprisingly, given human nature, the Web also has its dark side. It provides challenges that we must understand if we are to find a workable balance between two poles: (i) data ownership and privacy protection, (ii) freedom of information and intellectual property issues, and (iii) overwhelming surveillance and the free flow of information. For the Web to continue to grow as a force for good, we need to understand its societal challenges, including increased crime and socio-economic discrimination, and we must work towards fairness, social inclusion and open governance.

3.2 The Architecture of the Web

Before we can speak about the future of the Web, it is important that we understand what it is. This may sound easy, especially to anyone sitting in front of a laptop, tablet or phone reading this paper on their screen, but

DOI: 10.4324/9781003250470-4

in reality, the construct that we know as the Web is a fairly complicated, engineered object that requires explanation on a number of levels. As the construct, properly known as the 'World Wide Web' has grown from its original inception, there has been confusion between the Web and the underlying system that supports it, between the Web and the applications that it supports and between the technical construct of the Web and the 'social machines' that it has created via that construct.

To start with, in understanding the Web, it is important to differentiate it from the Internet on which it essentially sits. Although in today's parlance, the terms 'Internet' and 'Web' are often used interchangeably, they are really quite different things. The term 'Internet' (or sometimes just 'the Net') started out as a collective term for a number of different networking approaches being explored in the early days of computing. In the 1970s, several scientists, notably Vint Cerf and Bob Kahn, began exploring how to create a common set of protocols that would separate the details of how the bits (1s and 0s) that made up computer-to-computer communications could be shared without having to know the details of the specific mechanisms by which this sharing occurred. They proposed two protocols, one called the Internet Protocol (IP) since it would span the separate network approaches below it. The second protocol, the 'transmission control protocol' (TCP) provided a standard means by which data could be reliably transmitted between the machines providing control over the IP layer. The two protocols together came to be called TCP/IP and in 1983 were accepted as the international standard for computer-to-computer communication (see Figure 3.1).

The Internet supports many applications that were developed separately to run on top of these protocols, which includes email, voice over IP communications and a number of other applications that require the routing of information between the machines.

In 1989, Tim Berners-Lee, a researcher at the *Conseil Européen pour la Recherche Nucléaire* (CERN), the famous nuclear research organization, proposed a hypertext system that could be one of the applications that ran on top of the Internet. In 1990, a formal proposal was published, and approved, that suggested the creation of a program called 'WorldWideWeb' (originally one word). The program consisted of several components, including a protocol for naming the items on the Web, a protocol for how machines would find each other (using the Internet) and exchange information, and a language for how the exchanged information would be displayed in a program called a 'browser'. The actual names of these items – 'URL' for the item naming, 'HTTP' for the protocol for exchange of information, and 'HTML' for that display language – are the fundamental building blocks of the program that was released in 1991, renamed 'World Wide Web.'

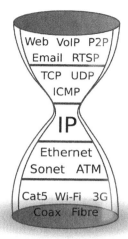

Figure 3.1 The 'internet hourglass' shows how the Internet Protocol separates higher level applications and protocols from lower-level networking.

In a very real sense, we could define the Web as no more than these three pieces (and their modern incarnations), as these are the fundamental building blocks on which the Web is based (cf. Berners-Lee et al. 1994).

One might make an analogy to the Internet being like the wires that connected homes to electricity, with the Web being the lightbulb that truly made that electricity useful. However, once homes had electricity and light, many other products that could use that electricity, and many other fixtures that made those lightbulbs either more useful or more attractive, were rapidly developed. Similarly, as the Web became more successful and entered more homes, augmentation came that blurred some of the definitions. Additionally, as the lightbulb changed the way people lived, making possible interactions that had been previously difficult or impossible, so too did the Web.

For example, one of the most important additions to the growing Web was the Mosaic browser released by the US National Center for Supercomputing Applications (NCSA) in 1993 (cf. Schatz and Hardin 1994). While not the first browser, Mosaic was easy to use, easily (and freely) available and required very little knowledge on the part of the user to be able to *surf* the Web (as its use came to be called). As people became more familiar with the Web and as browsers were commercialized and became more easily available for the emerging personal computer market, other uses came to extend the capabilities. For example, web logs (or blogs) started to grow and sites that aggregated the blogs also came along. In addition, as it became harder to find information on the growing web, new 'search sites' were developed that would help users to find the content they were looking for. The most successful of these was Google, which became, over the next decades, one of the most successful companies in the world, dominating a major share of web commerce.

In the late 1990s and early 2000s, user content on the Web started to really take off. New sites such as Flickr and the emerging social-networking sites (such as Myspace and, later, Facebook) allowed users to more easily share their own content. The movement to increase user capabilities to do this sharing took on the name 'Web 2.0' and many companies started to offer different kinds of information-sharing capabilities. The most successful of these companies was Facebook, which rapidly grew and, at the time this paper is being written, has billions of users. Another major change came around the same time with the advent of better sharing of video content. The website YouTube, released in 2005, was another major success and, as of 2017, it was reported that over one billion hours of content are watched on the site every day!

Along with the growth of the multiplicity of new web capabilities, there was also a growth in a new market – that of mobile computing. The release of the iPhone in 2007, and its incredible success, changed the way many users could access the Web. Instead of having to use a computer in a fixed location, as the number of users with mobile phones increased (whether an iPhone or a competitor's brand), a number of applications came along that could make the phone more useful. Many of these applications appeared to make the traditional browser no longer necessary for many users, and a confusion between what was and was not the Web began to grow.

The confusion is increased because some aspects of the original web, particularly the URL naming conventions and the HTTP protocol, remain critical to the success of these mobile applications. However, just as the Web hid the underlying details of the Internet, so too do these applications hide many of the details of the Web. For example, many users will now first open their phone to Facebook, Instagram, TikTok or some other social application. They can then click on an item in that application and the system will show them the content from a news site, a video, a photo, etc. The protocols used by these systems are the Web protocols (and others built on top of them), modified for mobile use. As such, many users now access web content through their phones or other mobile devices, rather than through a browser. A popular misconception has therefore arisen, which is that the information on these phone apps are somehow not 'the Web,' which is thought by many to be the specific application of the browser running on a personal computer or laptop.

The reason this matters, and why it is important in understanding the future of the Web, is the interconnectedness of these many applications. While these apps may seem disconnected to a user, information that may flow between different sites and applications can be as easily shared, as was the information linked between the hypertext pages of the early browser. Extensions to the web protocols have made tracking usage through this welter of applications easier for developers, while hiding many of these details from the users. It is these extensions that cause many of the privacy challenges of the Web, whether in a browser or on a mobile platform.

3.3 Privacy, Politics and the Centralization of the World Wide Web

Consider a typical day of life online – one might visit a news site to catch up on the day's headlines, look for recommendations on some products, purchase those products, check one's favorite social networks and perform many other day-to-day activities pursued either through a web browser or via mobile applications. While the user visits these sites, an extensive amount of information exchange is occurring in the background.

Figure 3.2 shows the 'trackers' – tools that keep track of users' activities on the Web, as they are invoked for several common websites. These trackers allow sites to keep tabs on who is visiting their sites, what other

Figure 3.2 A sample of the trackers used in (from left to right) a major news website, a popular travel planning site, an online site for buying shoes, an online book seller and an online dating site.

sites users may visit, what products or services users may be exploring, etc. As these systems collect this data, they also may have business relationships with each other, or with other entities, and thus buy and sell user information. Third-party companies, which can purchase the information collected through this tracking mechanism, are thus able to obtain significant amounts of user data.

When a user utilizes a mobile phone application to access online services, the amount of data collected can be significantly increased. The information tracking can be combined with geolocation data (where the user is), contextual information (what other applications may be open) and specific data about the user coming from the phone. This latter information, for example, can tell whether a user is walking vs. in a vehicle, whether the mobile device is being actively or passively engaged and even whether a user is inside vs. outside, and, if inside a building, phones can determine what floor of that building they are on. The ease with which this information is collected and correlated is made possible, in part, by the ways in which the applications are using the web protocols, described previously, to support the exchange of information.

The data collected can be used for many purposes. The most usual, of course, is targeted advertising, which can match users and ads (and provide a revenue stream to both the advertiser and the ad-matching company). Other uses of this information by corporations include tracking buying preferences to develop marketing campaigns for their products, sentiment tracking to better understand how their products are being received by customers and a wide range of other 'business intelligence' applications. This information also reveals much about the demographics of users and correlating that with political preferences has also led to tremendous use of this information by political campaigns and by governments that are monitoring, or trying to influence, elections.

The collection and use of data for advertising and marketing certainly predates the Web and has been used offline for election purposes and the like. The difference is that with the advent of the Web, and the scale of its usage, the amount of data that can be collected and linked is many orders of magnitude more than has been used in the past. Coupled with more powerful computers and better analytic tools, the data can be used to personalize the information collected well beyond what was doable in the past. The ability to determine the properties of a particular user, identified directly (for example, tied to a Facebook, Google, Baidu, WeChat, TikTok or other account), allows companies to specifically tailor and target services for a user or to direct specific messaging based on the user's previous interaction history.

As an example of the level of personalization this kind of information collection allows, differential pricing is enhanced to an individual level on the Web using this information. Thus, two users going to the same website, and looking at the price of the same product, may see different prices based on algorithms that use the collected information to determine the likelihood that the user will make the purchase. Similarly, a user returning to the same site, and looking at the same product, may see a higher price during the subsequent visit because the fact that they have returned may indicate an increased propensity to buy.

While companies argue that the collection and use of online data for advertising and similar purposes allows them to provide better service to their customers, on the one hand, and higher profits, on the other, the data can also have uses outside e-commerce. For example, it became clear during the 2016 elections in the United States that information collected from Facebook and other social networking sites was used to attempt to influence the election results.[3] Information collected by governments via CCD cameras and other devices that can be connected to online user data is increasingly being seen as a privacy threat on the one hand, while, on the other, governments argue for the need to use this information for law-enforcement, counter-terrorism and other such uses.

While the arguments as to the positive vs. negative use of the information continue, aspects of this argument take an effect not just on society, but on the Web itself. As the propensity for information use increases, there are several forces that challenge the very openness of the Web that allowed its unprecedented growth. Within the commerce sector, several companies that were able to become early innovators, and to build the data stores about users, have thus become, essentially, monopolies on the use of the Web. Companies such as Google, Amazon and Facebook in the US and Europe, or Baidu and Alibaba in China, have grown to the scale where the information they control has allowed them to become corporate monoliths that can control access and interaction on the Web. This makes it hard for new innovators to compete because their ability to rapidly amass the sorts of information stores that these larger companies have is highly constrained. The information collected on users also allowed large companies to offer the personalized services that keep the users within the 'walled gardens' that are their own sites and subsidiaries, again making innovation in the web marketplace more difficult.

A second, and much larger, threat to the Web comes from the desire of governments to control as much of the powerful information on the Web as possible and/or to deny access to the Web by others. For example, the Chinese government has long had a policy of blocking many sites from Chinese users. As time has gone on, the 'great firewall, as it has come to be known, has grown in scope and allows the government to control significant amounts of online information. In other countries, the ability of governments to censor or redirect web content, which was made difficult by the open design of the Web, has grown, allowing more control over web traffic by directly monitoring the information flows on the Web.

This 'splintering' of the Web, with increasing control of information by large companies on the one hand, and governments on the other, has been a force working against the openness of the Web.

3.4 The Future of the Web

A quote attributed to Alan Kay, then of Xerox PARC, is that 'the best way to predict the future is to invent it' and, for the first decade or so of the Web, that seemed to be a good model. Most of the key features of Berners-Lee's invention – the openness, the free exchange of information and the ability to scale – all were evident in the early web. As the Web entered its second decade, larger companies that invented or improved on the key technologies, such as search, writable web technologies (blogs, photo sharing, etc.), and social networking companies, seemed to be inventing the technologies that let us see how the Web would grow.

However, with the growth in complexity and the scale of the Web, the future became harder and harder to predict. As discussed above, the interaction of these created technologies as the companies became large monopolies have led to Web-based companies becoming some of the largest and most influential in the world. These companies now have a disproportionate control over the Web's growth and the power to suppress technologies that threaten their profits. The potential social impacts of the Web, and the desire of governments to control these, can also change how the Web can grow. Inventing new technologies at this scale becomes impossible and, thus, predicting the future can only be done by extrapolation from where we are today.

In an effort to explore possible futures, we present two contradictory futures, both based on trends we see starting today[4].

3.4.1 Scenario 1: The Dystopian Future

One clear possibility for the future of the Web is that the negative trends discussed so far take the lead in influencing the Web's future. Following this view, we see a number of trends coming together.

First, as companies that have grown larger and larger and control more personal information increasingly dominate Web use, they often also want to dominate access and suppress competition. A further challenge in this future is that internationalization, rather than becoming a unifying force, can become a separator. For example, currently, use of Google is prohibited in China.[5] While this resulted from a combination of political and economic factors, it allowed Baidu to grow to dominate online search in China (with over 75% of the market) and to become, by 2018, the fourth largest site on the Web in terms of Internet traffic. Baidu, like Google, is also competing in areas such as AI technology on and off the Web, autonomous vehicles and mapping and streaming TV, to name but a few.

Clearly, when giant companies like this are forming separate niches – one primarily focused on the English-speaking world and one aimed primarily at oriental languages – it makes the free flow of information, one of the original goals of the Web, much harder. Other search engines, such as Yandex, which focuses on providing services for Russian-language users, must compete directly with Google, which now spans many languages. Similarly, Baidu is expanding their engine to compete further in India and other emerging markets. While open markets and competition are often seen as good things for consumers, in these cases the monopoly power of the big companies in controlling access and content means that users in different countries, or using different products, are often directed to different answers – the search companies can control access to information, either directly or implicitly, by what they allow users to find. A similar phenomenon happens in social networking, where international giant Facebook competes with Chinese companies such as Tencent, which produces the WeChat application that dominates the Asian market, and in microblogging, where Twitter competes with Weibo, its Chinese equivalent.

It is worth noting that this competition between the large companies that are vested primarily in the major languages and largest economic blocs also disadvantage those speakers of other languages. For example, although Spanish is spoken throughout much of Central and South America, Google still dominates that market, with well over 90% of the market share. The content is multilingual, but the collection and crawling that determines the results is primarily dominated by algorithms developed for the US and European markets and thus contains implicit cultural biases.

While this sort of international competition is nothing new, it happens in many technologies, for example automobiles and mobile phones, and the impact of this splintering on the Web plays out in more areas than just the market competition. As alluded to above, the different providers produce different answers – some under the control of governments, some impacted by advertisers or other markets – which can influence how the people in those countries interpret news and events. Analogies are often made to the dystopian world of George Orwell's *1984,* where information is controlled by the government and free thought is discouraged.

This trend towards control of information and thought plays out in many other ways on today's Web and looks to be a trend that is increasing in intensity and variety. For example, political or religious groups are also trying to control web access or to build sites that can control content. The social media site 'Gab' in the US, for example, traffics in what is known as *alt-right* – often neo-Nazi content. Users are largely able to exchange information on their own views, without being challenged by others (and with the site's editors explicitly removing content that challenges their beliefs). This also plays out in other forms. For example, there are a number of religious social networking sites, many of which grew out of dating sites that were aimed at letting people find others of their own religion online. Sites that limit users based on gender preference, political and religious views, etc. also abound.

As well as these forces that cause content to be differentiated, there are also other activities online that are increasingly causing problems on the Web. In a 2018 article, web inventor Berners-Lee wrote:

> In recent years, we've seen governments engage in state-sponsored trolling to quash dissent and attack opposition. We've seen hacking and foreign interference distort politics and undermine elections. And we've seen how the spread of fake news on social media can trigger chaos, confusion and lethal violence.
>
> (Berners-Lee 2018)

These trends lead to a dystopian Web future, where people with different opinions tend to primarily receive information that is controlled by their governments, by giant Web companies or by the 'information bubbles' that let them see information only through the lenses of those information providers. They cater to people's political views, religious beliefs, or other preferences that limit the free flow of information and the very openness that made the Web grow in its first decades.

3.4.2 Scenario 2: A More Open Web

Human society, despite its many faults, has managed to deal with many of the issues above in surprisingly robust ways. Free markets, democratic systems and transparency of information have often mitigated against the worst attempts to control or censor information. Some of the darkest times in human history have been caused by wars between competing ideologies or religious beliefs, but these can be contrasted to times of growth and the flourishing of social consciousness.

It is not hard to see these positive trends also playing out on the Web. For example, a large movement in the scientific community is looking to scientific data sharing on the Web to create a more open practice of science to allow a better flourishing of research (cf. US National Academies of Sciences, Engineering and Medicine. 2018). The growth of crowd-sourcing sites and other kinds of 'social machines,' in which large numbers of people work together to provide a social good, are also important innovations on the Web. For example, Wikipedia has become a major source of information. This largely open and crowd-created encyclopedia has become the fifth most used site on the Internet (with over thirty-five million users of the English version reported to date). Other social machines include citizen science sites, crowdfunding sources, physical and mental health support sites and many others (cf. Hendler and Mulvehill 2016).

Just as there are governmental forces aiming to censor and control web content, there are others working to open it. Perhaps the best known of these on the current web is the EU's Genera Data Protection Regulation (GDPR), which was approved in 2016 and went into full effect in May of 2018. The GDPR was designed to 'protect and empower all EU citizens' data privacy' and to 'reshape the way organizations ... approach data privacy.'[6] In the US, while no such laws have yet been passed, there is a consideration of what has been called an 'Internet bill of rights'[7] that proposes six principles to ensure the protection of Web users:

- The right to universal Web access.
- The right to net neutrality.
- The right to be free from warrantless metadata collection.
- The right to disclose amount, nature and dates of secret government data requests.
- The right to be fully informed of scope of data use.
- The right to be informed when there is change of control over data.

While the chances of such a law passing in the US in the near future are low, over time it seems likely that some GDPR-like legislation will be passed. As more concern grows over information use and the issues of data ownership, it is likely that many governments around the world will be exploring their own versions of these kinds of restrictions.

As well as efforts within governments, Tim Berners-Lee is also leading an effort through the 'Web Foundation,' an organization he helped to found, called the 'contract for the Web.' A set of principles regarding what governments, large companies and individual citizens can do to let the Web live up to its original vision of openness are being developed. A set of 'starter' principles[8] have been proposed and a large group of organizations and individuals are working to develop key foundations, ranging from governments ensuring that everyone can connect to the Web to companies respecting consumers' privacy and personal data. Citizens are charged with being creators and collaborators on the Web, not just passive consumers, and are asked to help 'build strong communities that respect civil discourse and human dignity.'

In short, in contrast to the ownership and control models of the dystopian future, we see a more positive view, where the original principles of the Web as an open and free information sharing platform would be reasserted.

3.5 The Future of the Web Redux

The World Wide Web, as we have seen in this chapter, is a combination of an underlying internet of interconnected machines that enables the networking and exchange of information between people at a scale previously unimagined. In a world where it is estimated that as many as 60% of the population still does not have access to some basic health needs, such as adequate water sanitation, the reach of the Web into that same percentage of the population is a staggering achievement. Driven in part by access to the Web, world literacy has been increasing over the past decades and new models of what it means to be literate in an online world are being developed (cf. Leu et al. 2017).

While it is possible that the future will hold a dystopian Web that isolates people – or perhaps the more idealistic vision scene in the contract for the Web – the likelihood is that the truth will be somewhere in between, with features of both. Totalitarian and authoritarian governments are not likely to willingly give up their control of the Web. Neither is it likely that the large companies controlling so much of our access to information and collecting our personal data are likely to easily yield their sway. However, counterbalancing forces, including legal remedies, such as the GDPR, and social forces, such as the use of the Web for societal benefit, mitigate against the worst of these trends. Given the Web's incredible scale, and its underlying complexity predicting how the future will evolve is extremely difficult. However, comprehending the Web's technologies, exploring its social impact and examining how these two interact is crucial to being able to understand the effects that this increasingly universal information space is having, and will continue to have, on the future of our world (cf. Hendler and Hall 2016).

Notes

1 Available at https://www.itu.int/dms_pub/itu-s/opb/pol/S-POL-BROADBAND.19-2018-PDF-E.pdf Accessed at 15.11.2021.
2 Available at https://www.statista.com/statistics/617136/digital-population-worldwide/ Accessed at 15.11.2021.
3 The extent to which the online use of this data actually influenced the outcome of the election is still being debated, but clearly the ability to collect and use this information was considered to be a significant factor (cf. Berghel 2018). The Brexit vote in Britain, and several other elections, were also influenced in this way. Available at https://www.politico.eu/article/cambridge-analytica-chris-wylie-brexit-trump-britain-data-protection-privacy-facebook/ Accessed at 15.11.2021.
4 A detailed analysis of possible futures for the Internet, as opposed to the Web, is presented in O'Hara and Hall 2020.
5 There is a good discussion of Google's history in China on Wikipedia, available at https://en.wikipedia.org/wiki/Google_China Accessed at 15.11.2021. See also Roberts 2018.
6 Available at https://eugdpr.org/ Accessed at 15.12.2018.
7 Available at https://www.rokhanna.com/issues/internet-bill-rights Accessed at 15.12.2018.
8 Available at https://fortheweb.webfoundation.org/principles-1 Accessed at 15.11.2021.

References

Berghel, H. 2018. 'Malice Domestic: The Cambridge Analytica Dystopia', in *Computer*, vol. 5, pp. 84–89.

Berners-Lee, T. 2018. 'How to Save the Web. Turning Points Column', in *New York Times*, Dec 6. Available at https://www.nytimes.com/2018/12/06/opinion/tim-berners-lee-saving-the-internet.html Accessed at 6 December 2018.

Berners-Lee, T., Cailliau, R., Luotonen, A., Nielsen, H. and Secret, A. 1994. 'The World Wide Web', in *Communications of the ACM*, vol. 37/8, pp. 907–912.

Hendler, J. and Hall, W. 2016. 'Science of the World Wide Web', in *Science*, vol. 354/6313, pp. 703–704.

Hendler, J. and Mulvehill, A. 2016. *Social Machines: The Coming Collision of Artificial Intelligence, Social Networks and Humanity*, New York: Apress.

Leu, D. J., Kinzer, C. K., Coiro, J., Castek, J. and Henry, L. A. 2017. 'New Literacies: A Dual-Level Theory of the Changing Nature of Literacy, Instruction, and Assessment', in *Journal of Education*, vol. 197/2, pp. 1–18.

O'Hara, K. and Hall, W. 2020. 'Four Internets', in *Communications of the ACM*, vol. 63/3, pp. 20–30.

Roberts, M. E. 2018. *Censored: Distraction and Diversion Inside Chinas Great Firewall*, Princeton, NJ: Princeton University Press.

Schatz, B. and Hardin, J. 1994. 'NCSA Mosaic and the World Wide Web: Global Hypermedia Protocols for the Internet', in *Science* vol. 265/5174, pp. 895–901.

US National Academies of Sciences, Engineering and Medicine. 2018. *Open Science by Design: Realizing a Vision for 21st Century Research*, July 2018. Available at https://www.nap.edu/catalog/25116/open-science-by-design-realizing-a-vision-for-21st-century Accessed at 15 December 2018.

Part II

Structural and Dynamic Complexity in the Design of the Internet

4 Designing an Internet of Machines and Humans for the Future

Yan Luo

4.1 Introduction to the New Internet of Machines and Humans

The Internet has become the foundation of information technology of human society since its inception in the 1970s. Its first operational form, ARPANET (an interconnection of academic and military networks), despite the excitement it generated among the researchers and engineers, did not reveal at the onset how it would fundamentally change the way people interact and share information in the next century as it does now. The Internet today is a platform for information acquisition, a catalyst for information fusion and a driver for ever-enhanced machine intelligence, serving the human civilization in the new digital age.

The ubiquity of new sensors has made it indispensable to gather environmental information for a plethora of emerging applications, ranging from smart homes to autonomous driving. The temperature and humidity in a home can be monitored continuously and adjusted to suit personal preferences. The emergence of autonomous driving is built on top of high-definition cameras and distance-sensing LIDAR equipped on a car that can perceive the road conditions in real time. The health-oriented wearables, such as smart watches, detect heartbeat changes and body motion to provide wellness recommendations. The public generally welcome these sensory devices, however they become increasingly cautious with the privacy and security of the data gathered, demanding strict regulations such as HIPAA (HHS 2013) and GDPR (GDPR 2016).

The functionality of data transferring remains the root of the Internet, whose responsibilities, however, have evolved to a new level beyond simply moving bytes. The privacy and integrity of data, fine-grained access control, the verifiable transactions of commerce and many new requirements on information exchange have imposed so many new challenges to the Internet infrastructure that it constantly embraces new enhancements, such as differential privacy (Dwork et al. 2006), software defined networking (Kim and Feamster 2013), blockchain (Nakamoto 2009; Wood 2021), 5G infrastructure (Gupta and Jha 2015) and, ironically, new limitations soon afterwards.

DOI: 10.4324/9781003250470-6

One of the largest technological leaps in the past decade is the resurgence of machine intelligence. The massive computational capabilities in graphics processing units (GPUs) are now able to effectively cope with the complexity of deep neural networks, which are the foundations of recent advancement in object classification, human pose recognition, tracking and analysis, robotics, medical imaging diagnosis, Go chess championships and so on. It is the Internet that brings such machine intelligence from the cloud computing facilities to the end users and drives novel applications, benefiting the fields of medicine, transportation, manufacturing, sports and so on.

In this chapter, we attempt to provide a holistic overview of the building components and drivers in the current Internet, an in-depth analysis on how these components influence each other and a bold prediction on the design trend of future Internet in the next decade. In such a way, we expect to shine some light on the design philosophy of future Internet – which academic researchers and industrial practitioners can verify, criticize, complement and amend – and to build together an ever-evolving, useful, sustainable network of machines and humans.

This chapter is organized as follows: Section 4.2 provides a background; Section 4.3 gives an overview of the current Internet design; Section 4.4 outlines some predictions on the future design of the Internet; and Section 4.5 expresses some relevant conclusions.

4.2 Background and Related Work

Yin et al. point out that big data drives the new design of future Internet in architecture, services and applications. The authors envision that in the future Internet: '(1) computational complexity replaces state complexity in the control plane; (2) data intelligence enables user choices and rewards innovations; and (3) correlations from data analytics help solve inherently hard optimization problems' (Yin et al., 2014). The predictions are centered around the metrics regarding the Internet itself, such as network delays and the user's quality of experiences, collected by network probes and web applications, which are important sources of large volumes of network data. This work is the first one to highlight the trend of data-driven network design and control, and it mainly focuses on the architecture, communication models, resource management and future of the Internet. In this paper, we aim to provide an overview of the recent technological advancement in many fields closely related to the Internet, and rethink how the Internet will evolve and how we should design it with the lessons learned so far.

4.2.1 *The Internet of Things*

The concept of the Internet of Things (IoT) has come a long way since Mark Weiser's 1991 paper on ubiquitous computing, 'The Computer of the 21st Century' (Weiser 1991). Recent academic venues such as UbiComp and PerCom yield the contemporary vision of the IoT, which range from personal wearables to sensors in cars, buildings, etc., connected through the Internet. From a system architecture perspective, an IoT device consists of sensors, processors and networking units. IoT devices and their functionalities are made possible by the energy efficient computing architecture and algorithms on today's embedded hardware platforms (Hester et al. 2016).

Wearables are the tangible devices designed for monitoring a person's vital signs and body motion. Examples of such are Apple Watches and Fitbit activity trackers, which can continuously obtain sensory data from accelerometers, gyroscopes and magnetometers to classify and assess activities (e.g., walking and sleeping). Some devices can also gather heartbeat or even electrocardiogram (ECG) information to accurately monitor the health of human hearts. Wearables are not limited to being worn on the wrist anymore. For example, new technologies can now mass-produce fabric with embedded soft and stretchy sensors for monitoring finger motions (Atalay et al. 2017).

In general, sensors are deployed far more widely than wearables. In manufacturing plants, sensors are utilized to measure vibration, overheating and freezing, movement or proximity of objects, status of valves, buttons and switches, concentration of chemical particles, etc. in a wide range of industrial environments. These devices are critical for the safe and continuous operation of machines and smooth production procedures. In cars, sensors are installed to monitor tire pressure, temperature of coolant as well as engine, friction of brakes and distance from the car in front, which assist the driver in maneuvering the car safely on the roads. In hospitals, a variety of sensors are used in oxygen concentrators, sleep apnea machines, kidney dialysis machines, infusion and insulin pumps, blood analyzers, respiratory monitoring equipment and blood pressure monitoring equipment to deliver the proper medical care for patients.

Among all the sensors, cameras have become increasingly popular and important in surveillance, manufacturing and healthcare to provide rich information about subjects and their surrounding environment. The vast amount of video data has introduced interesting research problems in digital image processing, pattern recognition, computing and networking, which those academic has been investigating since the 1980s when computing cost started declining. The focus of image processing started from filtering (Li et al. 2015b) (transforming the original image into its new representation) to object detection (Girshick et al. 2014) and now has shifted to object classification, subject tracking and scene analysis (Li et al. 2015a).

The increasingly significant impact of IoT is due to the network connectivity of these devices, which enables advanced data fusion and analytics when the sensory data are collected in large quantities and over a significant period of time. The sensor data are transferred from the device to a data aggregation point (a smartphone, a computer or a cloud service) for analysis and correlation, to understand the status of a person, a machine, or the environment. Such knowledge then drives decisions surrounding the control of a physical cyber system. Without the network connectivity, the sensor data would be much less valuable, as they are isolated and do not provide knowledge of the human or machine being monitored. Even with connectivity, the sensor data, if not transferred in time, will become stale and must be discarded. Therefore, the center of the IoT remains the Internet itself, which has gone through dramatic changes in the past decade.

4.2.2 *Programmable Networks*

The Internet consists of personal networks and enterprise networks, which are connected through Internet Service Providers (ISPs) and their network backbones. The global scale of the network architecture has been built on a set of network protocols (e.g., TCP/IP, BGP) and heterogeneous network equipment, which run these protocols. There are two major issues with the conventional network architecture: one is heterogeneity and the other one is the lack of automation.

Why is heterogeneity bad? Heterogeneity comes from the variety of network standards, many network equipment vendors and their clear product differentiation. But the network protocols in fact mandate homogeneity and standards. The heterogeneous network equipment and their management interfaces have introduced increasing complexity in network management and control, not only across organizations and domains, but also within an organization. Complexity leads to increased costs and potential errors. The network faults caused by the complexity of equipment management and associated human errors constitutes about 50–80% of the network outages (Juniper 2008). The consequences of network downtime are more than financial in nature. They can be fatal in critical missions.

Why is automation important? As the networks scale in size, the number of network devices in an enterprise network can be in the order of hundreds, while that number can go beyond thousands and even millions in a large data center of cloud service providers. To individually configure every device requires infeasible human effort, therefore an automated management flow and other associated tools are necessities. Automation also reduces the chance of human errors if proper validation schemes are in place.

Separating a network control plane from its data plane address and defining an abstraction of the control plane with Application Programming Interfaces (APIs) elegantly addresses both issues of heterogeneity and automation. The idea of separation of control and data plane provides a chance to rethink the fundamental functionality of network equipment, which is to match the incoming network datagrams (packets) with preset criteria and to determine the actions on them (e.g., move forward to the next device, or drop the packets). The essence of the abstraction is the 'match-action' primitive, with which existing network functions such as routing, firewalls and load balancing can be realized. OpenFlow (McKeown et al. 2008) as the outcome of such separation and abstraction has become one of the most exciting innovations in networking, by bringing software algorithms, verification methods and system tools to the networking domain. This innovation has vigorously energized many new research initiatives and academic conferences (ACM 2015) and has led to a new billion-dollar software-driven network industry (Markets and Markets 2020).

It is worth noting that OpenFlow has become the synonym of Software Defined Networking (SDN), which has its far-reaching impact on the network data plane as well. This network data plane refers to the fast path across which a network packet traverses a network device. The data plane has long prioritized performance (gigabits per second) over flexibility and programmability. As a result, the data plane has been rigid; Any changes in packet format or processing methods are out of the question because such changes will immediately make the existing hardware non-compliant or non-operational. This rigidity has been a major hurdle that impedes innovations in network protocols and applications. To address this problem, SDN goes further with P4 (Bosshart et al. 2014), a protocol and target independent domain of specific language, and brings the programmability into a data plane. With P4, one can describe their own way of parsing the header and content of a network packet, apply actions such as prefix matching on flexibly chosen packet header fields and even rewrite packet headers, if needed. This unprecedented programmability in a network data plane level enables the introduction of new protocols, new packet formats and new processing methods at a much more rapid pace.

SDN has brought deep programmability to the Internet architecture. Such programmability empowers people with tools that make network management more flexible, resilient and scalable. With such programmability, the Internet is now in a better position to connect various resources to devices, applications and people in an automatic, predictable and responsive fashion.

4.2.3 Machine Intelligence

Machine intelligence has come a long way from its infancy to today's status. Turing's seminal paper 'Computing Machinery and Intelligence' in 1950 laid out several criteria to test whether a machine has intelligence, known as the 'Turing test' (Turing 1950). The paper ignited a monumental amount of research on Artificial Intelligence or machine learning which has centered around the theory of how human brains work, modeled as neural networks. One of the most notable works is Rosenblatt's 'perceptron' (Rosenblatt 1958), the foundation of modern neural networks. The essence of perceptron is a learning procedure that would probably converge to the correct solution and could recognize images, as demonstrated with a custom-built hardware called 'Mark 1 perceptron'. More than a decade later, in the 1969 book *Perceptrons*, Minsky and Papert questioned the capabilities of perceptrons and attempted to provide a mathematical framework for understanding the performance of perceptrons in general. However, the book unfortunately made the perceptron style machine intelligence unfashionable until the mid-1980s, partly because the computational requirements far exceeded what was available. In 1986, Rumelhart et al. showed in a paper entitled 'Learning Representations by Back-propagating Errors' that neural nets with many hidden layers could be effectively trained by a relatively simple procedure. This liberates neural nets to cope with the weakness of the perceptron because the additional layers furnish the network with the ability to learn nonlinear functions. Their proposal of 'backprop,' taking the derivative of the network's loss function and back-propagating the errors to update the parameters in the lower layers, is a significant step that later led to the success of training convolutional neural networks (CNNs).

Hinton et al. presented the method of training a simple two-layer unsupervised model, freezing all the parameters and then training just the parameters for the new layer added on top of the previous layer (Hinton et al., 2006). The resulting parameters can be used in the final neural network. This 2006 paper marked the beginning of the famous 'deep learning' era. The breakthrough year of 2012 for deep learning witnessed the Large Scale Visual Recognition Challenge using a very large image database, 'ImageNet', with millions of labeled images (Deng et al. 2009). The popularity of the contest generated tremendous interest in deep learning. Since then, the research and practice of deep neural networks have generated a vast amount of knowledge of the neural network architecture, training and optimization, as well as their applications in computer vision, natural language processing, human-computer interaction and many other domains.

The microprocessor technologies' advancement further propels the machine intelligence leap by providing high-performance computing to solve real-world problems effectively. A typical deep learning network such as AlexNet consists of five convolutional layers, some of which are followed by max-pooling layers, and three fully-connected layers with a final 1000-way softmax (Krizhevsky et al. 2012). This deep learning network attached ReLU after every convolutional and fully-connected layer. To achieve the best error rate at the time of its publication (top-1 and top-5 error rates of 37.5% and 17.0%, respectively), AlexNet was trained for 6 days on two Nvidia Geforce GTX 580 GPUs. As the training data proliferate and neural network models grow in complexity, the computational requirement of deep learning tasks are so tremendous that the machine intelligence can only be achieved in the data centers of cloud computing with abundant high-performance processors (e.g., GPUs).

4.2.4 Privacy and Security

The first areas to be considered regarding privacy and security are cloud computing and cloud storage. Despite the great success of cloud computing, its security and privacy are still the primary concern, hindering its wide adoption, especially for those enterprises and organizations that have a high demand on data security and privacy.

Data outsourcing, as the key application of cloud computing, enables individuals and companies to save costs on infrastructure construction and relieves them of the heavy burden of data management. However, data outsourcing actually hands over owners' control of data to service providers, which potentially brings many new security threats. The most essential reason is that cloud-based applications involve at least two independent security domains and no domain will put full trust in another, which makes the security and privacy problem in cloud scenarios more complicated. During the past decade, many noteworthy cloud security breaches have appeared (StorageCraft 2021). Multi-tenancy is another important characteristic of cloud computing, which allows cloud providers to manage computation and storage resource in an economic and efficient way, by providing virtualized and shared infrastructure resource to customers. Thereby, the providers should guarantee strong isolation and mediated sharing between VMs to protect multiple tenants' data (Wei et al. 2009).

The Internet of Things, Big Data and Edge Computing are interrelated. Besides inheriting the same security issues from RFID systems, sensor networks, mobile communication networks and the Internet, the IoT has its own specialties in security and privacy, such as authentication, access control configurations, secure information storage and management, etc. Furthermore, due to the heterogeneous architecture of the IoT, it should address compatibility issues between different networks, which makes it prone to many security threats and attacks, e.g., denial-of-service and distributed denial-of-service attacks, forgery/middle attacks (Roman et al. 2013).

An IoT system is an integration of different layers in its architecture (Shi et al. 2016). Hence, IoT security and privacy should address security problems in these layers: the perception layer needs to address RFID security, WSNs (Wireless Sensor Networks) security and RSN (Robust Security Network) security; the transport layer needs to deal with the security of involved 3G/4G access networks, ad-hoc networks and Wi-Fi networks; the application layer involves middleware security and cloud platform security.

As millions of devices get connected, the IoT triggers a massive inflow of big data, increasing at a rapid rate. Among them, a large portion of data are produced at the edge of the network. It would be more efficient to process these data instantly because it would be inefficient to transfer such data to the cloud for processing, considering the massive bandwidth it would consume. Edge computing can mitigate the data transfer and process burden of cloud platforms by processing data at the network edge. However, the complexity and heterogeneity of various edge devices and involved networks make its security and privacy a big challenge. Moreover, many devices are resource constrained, so they require the emergence of lightweight security and privacy preservation mechanisms. Finally, ownership and stewardship issues of the data collected from various edge devices should be seriously considered when processing these data.

Big data privacy should address privacy on data collection, storage and processing phases. Currently, there are many research attempts to address this problem, including homomorphic encryption (Gentry 2009), search over encrypted data (Li et al. 2013) and anonymity techniques (Sweeney 2002) to avoid identity disclosure.

4.3 Lessons Learned from the Design of Today's Internet

The current Internet is a complex system where the IoT, network architecture and intelligent services interact, support and complement one another. The making of the Internet is a result of technological advancement and policy compromises, driven by people's constant desires for better applications and services. The technologies are pursued to address the unique characteristics and usage patterns of the data, infrastructure and applications. In this section, we describe the nature of the building blocks of the current Internet broadly and summarize the lessons learned from this design process.

4.3.1 Characteristics of Data

We hereby define data as the samples outputted from IoT devices – they will generate more than 500 zettabytes per year in data and that number is expected to grow exponentially, not linearly (Cisco 2018). The majority of the services and applications on the Internet handle these data directly or indirectly.

Data generated from IoT devices are unstructured, random and enormous. The types of sensors and the format of the sensor data from the IoT devices can vary from numerical values (e.g., temperature) to wave forms (e.g., audio), to pixels (e.g., video). Although a single sensor tends to maintain a structured format, the data collected from a variety of IoT devices are unstructured and can be lossy. Some of the IoT devices run on a constant sampling rate and thus produce periodic data samples. However, a large portion of the sensors react to triggering events, therefore delivering data at random moments. The volume of sensor data is so large that consumers of these data require adequate processing capabilities to filter samples and carry out analytics tasks, either at the edge (close to the data source) or in the cloud.

Data are time sensitive. The purpose of the IoT is to have a manageable framework to monitor subjects or environments, support decision-making and control physical cyber systems. The feedback loop requires timely information collection. A decision made based on the sensor data would be problematic if the data are out of date or stale. Minimizing the delays in data retrieval, network transfer and processing is of great importance.

The requirements on data processing are diverse. Medical applications are governed by the HIPAA regulations, thus requiring the security and privacy of medical records and personally identifiable information (PII). The control system of an autonomous vehicle demands the shortest data processing delay to react to road hazards and avoid accidents. Analyzing surveillance videos requires the highest throughput (frames per second) possible to spot a suspect from the vast number of pedestrians who have walked through an intersection. Privacy sensitive crowdsourcing applications employ statistical techniques such as differential privacy (Dwork et al. 2006) to add a carefully tuned amount of random noise to raw data sets to hide identifiable information, while supporting close to accurate queries (Li and Miklau 2012). The underlying network has to support all these applications with different requirements and expectations, which has imposed interesting challenges to network architecture.

4.3.2 Characteristics of Network Architecture

The performance of networks keeps improving in terms of throughput, latency and reliability. The speed of networks has grown exponentially for both wired and wireless networks every year within the past decade (Zhuang et al. 2013). The download speeds of 5G mobile networks can reach at least 1 Gbps, which is equivalent to the speed of a typical wired network to an office computer today. The theoretical latency of 5G networks is 1 ms, 50 times shorter than the current 4G/LTE networks. Such a performance boost enables future resource-intensive applications such as remote medical imaging and surgery, self-driving cars, 3D immersive video collaboration and virtual reality.

The networks gradually gain the support of flexibility and programmability. The changes in network states are realized with software-based controllers executing algorithms, designed and verified based on network policies. The networking domain, along with computing and storage domains, is embracing a paradigm shift from proprietary devices to open, software-defined devices, APIs and services. To establish and manage a network effectively, the software skills (development and debugging) of a network administrator are now even more important than the proficiency of vendor-specific tools, which was the professional qualification prior to the SDN era.

The security and integrity of data transfer are increasingly important. The Internet has been employing encryption schemes such as IPsec and TLS at different network layers. The programmability of the network allows for flow level differentiation and an added layer of data security. For example, the network can enforce a policy that requires the data coming out of a medical device to be encrypted with TLS. Any violation of such a policy can be detected in real time at an OpenFlow-compliant network switch, which forwards to the SDN controller the first packet of every new data flow originating from the medical device.

4.3.3 Characteristics of Machine Intelligence

The labeled data used for training are the fueling substrate of machine intelligence. Deep learning, the center of today's machine intelligence, relies heavily on deeply layered neural networks. These neural networks count on labeled data to train the parameters of the network. ImageNet (Deng et al. 2009) is one such example, where over 15 million high-resolution images belonging to roughly 22,000 categories were collected and labeled by human labelers, using Amazon's Mechanical Turk crowdsourcing tool. Google uses quantitative evaluations with the Bilingual Evaluation Understudy (BLEU), a metric used in machine translation to evaluate the quality of generated sentences (Vinyals et al. 2015). As a side benefit, every Captcha image question, which a human answers by validating whether an image is showing a certain scene, helps the AI engine to train the models and eventually produce a natural language description of an image.

After the huge success of applying deep learning methods to solve real-world problems, machine intelligence powered by deep neural networks keeps growing and being validated with the new data and increased computation power. In general, there are two categories of neural networks: supervised and unsupervised. The popular Convolutional Neural Networks (CNNs) are an example of supervised deep learning (DL), while the emerging reinforcement learning (RL) is in the realm of being unsupervised. An architecture of the neural networks may comprise a variety of 'layers' such as convolutional, back-propagation, max-pooling, etc., each of which may repeat a number of times and interconnect with other layers to form a multi-layer network. The theoretical foundation of deep learning (TFDL) has recently gained a growing interest to systematically evaluate competing intuitions and hopefully lead to new insights and concepts. Towards this goal, the TFDL workshop (TFDL 2018) on the theoretical foundation of deep learning attracted researchers and practitioners in this field to brainstorm on 'generalization ability of deep learning, regularization schemes, adversarial training, generative models, training neural networks, and optimization.'

The computational demands from machine intelligence are constantly growing. The amount of computational power used in the largest AI training runs has been increasing exponentially with a 3.5 month-doubling time since 2002 (OpenAI, 2018). In contrast, Moore's law, which predicts the advancement of microprocessor technology, had an 18-month doubling period. Such computation power can only be harnessed from cloud based platforms such as Google Cloud, Amazon AWS and Microsoft Azure.

4.3.4 *The Making of Today's Internet*

The IoT, networking and machine intelligence are intertwined and influencing each other in their evolution processes.

IoT devices gain their importance mainly because of their networking capabilities to deliver sensory data continuously to analytics processes. In the past, sensors were utilized to assess the surrounding situations and only applied in localized decision processes. Low power embedded networking technologies provide sufficient communication distance and cost-effective hardware to sensors. With networking connectivity, the IoT changes the volume and timeliness of data, which, in turn, motivates wider deployment and new innovations.

The machine intelligence relies on the training data collected from the IoT devices. The gigantic amount of labeled data has made it possible to iteratively tune the parameters in neural networks to yield models that are effective in classification, tracking, speech and learning trends. To continuously improve the accuracy of machine learning models, new IoT devices are designed and deployed to gather in a larger quantity and more fine-grained data about humans and environment.

The Internet bridges the computing resources at the cloud with the data source at the edge and delivers the intelligence to users. To fulfill the requirements of new application scenarios and regulations, the network infrastructure gains programmability, resilience, privacy-preserving techniques and improved speeds and reliability.

4.4 Designing the Future Internet

Forecasting the future is always a speculation. In this section, we attempt to make a few predictions on how the design of the future Internet will be and initiate open discussions.

4.4.1 Deeper Programmability and Customizability

The Internet infrastructure will become divergent and deeply programmable in the next decade. The network infrastructure is expected to grow substantially in two directions: mobile networking and cloud networking. The future mobile networks such as 5G provide an order of magnitude higher throughput and more than an order of magnitude shorter latency. The design of future Internet will strive to enable bandwidth hungry and latency sensitive applications in the consumer, manufacturing and medical domains. The cellular coverage is smaller but denser, so smart algorithms are needed to allocate spectrum resources using cognitive radios (Mishra et al. 2006), a salient example of deep programmability. Likewise, cloud computing delivers important services for consumers and businesses, which gradually shift computational workloads from personal computers to shared cloud computing resources. As the responsibilities of the cloud expand, cloud services call for the reliability, security and programmability of the future network infrastructure.

4.4.2 Conduit of Computing Power

The Internet will become the conduit of computing power. As computing capabilities become dynamic in spatial dimension due to software-defined infrastructure and virtual machines, we are given an unprecedented choice to utilize the resources in a more cost-effective way. The computational demands from machine intelligence persist and are only expected to become larger, so it has been the trend to train complex neural networks in the cloud. At the same time, the aggregation of IoT data in the cloud with managed services and APIs allows centralized analytics on demand. We can afford to have thin mobile devices to generate content and consume rich information, while relying on the services run on the cloud. The functionality as a computational power conduit will be vested onto the future Internet.

4.4.3 Human Centered Internet

Humans have become the center of Internet evolutions and will remain so as the Internet is reshaped and enriched with novel designs of its infrastructure and its applications running atop it. The pervasive IoT sensors generate a massive amount of data about human themselves, and their environment, and most Internet applications leverage such data, analyze them, apply machine intelligence, and assist in controlling machinery, issuing timely alarms, providing insightful knowledge and so on. As human needs evolve in healthcare, manufacturing and environmental conservation, the future Internet is charged with such missions as to respond to the needs and challenges.

4.4.4 Reaching the Underserved

The access to Internet has never been equal geographically and socioeconomically, leading to large disparity between regions and social groups. However, the access to Internet and the data and computational resources provided by the Internet would largely impact on the likelihood of success and sustainable development of one's socioeconomic status. Therefore, one of the major objectives of the future Internet is to ensure that the inequality of access is eliminated, and that the Internet reaches as many people as possible, regardless their geographical location, income level, language or education background. Fortunately, the advancement of new technologies such as 5G and satellite broadband Internet have provided technological feasibilities, yet the stake holders (community, non-profit and governments) shall invest both capital and human resources to ensure the future Internet reaches the underserved.

4.5 Conclusion

It is the Internet that bridges the computational platforms in the cloud with the data source at the edge and delivers the intelligence to its beneficiary, the humans. With the rapid evolution of networking, computing and storage technologies, the paradigm of information technology has shifted from a personal computing era to an information society, where ownership of network connectivity outweighs the ownership of private computing facilities. The network architecture embraces divergent progression in both mobile and cloud networks and will become the conduit of computational power. Overall, the future Internet is charged with the tasks of amassing and transferring data as well as delivering scalable computational power to drive machine intelligence, meeting all the new requirements on performance, flexibility, security, privacy and reliability.

Acknowledgment

This work is supported in part by the U.S. National Science Foundation (No. 1450996, No. 1541434, No. 1547428 and No. 1738965). Dr. Hao Jin contributed to Section 4.2.4 (security and privacy).

References

ACM 2015. *ACM SIGCOMM Symposium on SDN Research* (SOSR). Available at: http://www.sigcomm.org/events/SOSR Accessed at 19.11.2021.

Atalay, A., Sanchez, V., Atalay, O., Vogt, D. M., Haufe, F., Wood, R. J. and Walsh, C. J. 2017. 'Batch Fabrication of Customizable Silicone-Textile Composite Capacitive Strain Sensors for Human Motion Tracking', in *Advanced Materials Technologies*, vol. 2/9, 1700136.

Bosshart, P., Daly, D., Gibb, G., Izzard, M., McKeown, N., Rexford, J., Schlesinger, C., Talayco, D., Vahdat, A., Varghese, G. and Walker, D. 2014. 'P4: Programming Protocol-Independent Packet Processors', in *SIGCOMM Computer Communication Review*, vol. 44/3, pp. 87–95.

Cisco. 2018. *Cisco Global Cloud Index: Forecast and Methodology, 2016–2021*. Available at: https://www.cisco.com/c/en/us/solutions/collateral/service-provider/global-cloud-index-gci/white-paper-c11-738085.pdf Accessed at 19.11.2021.

Deng, J., Dong, W., Socher, R., Li, L.-J., Li, K. and Fei-Fei, L. 2009. 'Imagenet: A Large-Scale Hierarchical Image Database', in *IEEE Conference on Computer Vision and Pattern Recognition, 2009 (CVPR 2009)*, pp. 248–255.

Dwork, C., McSherry, F., Nissim, K. and Smith, A. 2006. 'Calibrating Noise to Sensitivity in Private Data Analysis', in S. Halevi and T. Rabin (eds.), *Theory of Cryptography, Volume 3876 of Lecture Notes in Computer Science*, Berlin: Springer, pp. 265–284.

GDPR 2016. *General Data Protection Regulation*. Available at: https://gdpr.eu Accessed at 19.11.2021.

Gentry, C. 2009. *A Fully Homomorphic Encryption Scheme*. Ph.D. thesis, Stanford, CA: Stanford University.

Girshick, R., Donahue, J., Darrell, T. and Malik, J. 2014. 'Rich Feature Hierarchies for Accurate Object Detection and Semantic Segmentation', in *Proceedings of the IEEE Conference on Computer Vision and Pattern Recognition*, IEEE Computer Society, Washington DC, pp. 580–587.

Gupta, A. and Jha, R. K. 2015. 'A Survey of 5g Network: Architecture and Emerging Technologies', in *IEEE Access*, vol. 3, pp. 1206–1232.

Hester, J., Peters, T., Yun, T., Peterson, R., Skinner, J., Golla, B., Storer, K., Hearndon, S., Freeman, K., Lord, S., Halter, R., Kotz, D. and Sorber, J. 2016. 'Amulet: An Energy-Efficient, Multi-Application Wearable Platform', in *Proceedings of the 14th ACM Conference on Embedded Network Sensor Systems, SenSys'16*, November 2016, New York, NY: ACM, pp. 216–229.

HHS 2013. *Summary of the HIPAA Security Rule*. Available at: https://www.hhs.gov/hipaa/for-professionals/security/laws-regulations/index.html Accessed at 19.11.2021.

Hinton, G. E., Osindero, S. and The, Y.-W. 2006. 'A Fast Learning Algorithm for Deep Belief Nets', in *Neural Computing*, vol. 18/7, pp. 1527–1554.

Juniper 2008. *What Is Behind Network Downtime? Proactive Steps to Reduce Human Error and Improve Availability of Networks*. Available at: https://cdn2.hubspot.net/hubfs/2581329/Tellabs_Feb2017%20Theme/pdf%20files/200249.pdf Accessed at 19.11.2021.

Kim, H. and Feamster, N. 2013. 'Improving Network Management with Software Defined Networking', in *IEEE Communications Magazine*, vol. 51/2, pp. 114–119.

Krizhevsky, A., Sutskever, I. and Hinton, G. E. 2012. 'Imagenet Classification with Deep Convolutional Neural Networks', in *Advances in Neural Information Processing Systems*, vol. 25, pp. 1097–1105.

Li, C. and G. Miklau. 2012. 'An Adaptive Mechanism for Accurate Query Answering under Differential Privacy', in *Proceedings of the VLDB Endowment*, vol. 5/6, pp. 514–525.

Li, M., Yu, S., Ren, K., Lou, W. and Hou, Y. T. 2013. 'Toward Privacy-Assured and Searchable Cloud Data Storage Services', in *IEEE Network*, vol. 27/4, pp. 56–62.

Li, T., Chang, H., Wang, M., Ni, B., Hong, R. and Yan, S. 2015a. 'Crowded Scene Analysis: A Survey', in *IEEE Transactions on Circuits and Systems for Video Technology*, vol. 25/3, pp. 367–386.

Li, Z., Zheng, J., Zhu, Z., Yao, W. and Wu, S. 2015b. 'Weighted Guided Image Filtering', in *IEEE Transactions on Image Processing*, vol. 24/1, pp. 120–129.

Markets and Markets 2020. *Software-Defined Networking Market by Component (SDN Infrastructure, Software, and Services), SDN Type (Open SDN, SDN via Overlay, and SDN via API), End User, Organization Size, Enterprise Vertical, and Region – Global Forecast to 2025*. Available at: https://www.marketsandmarkets.com/Market-Reports/software-defined-networking-sdn-market-655.html Accessed at 19.11.2021.

McKeown, N., Anderson, T., Balakrishnan, H., Parulkar, G., Peterson, L., Rexford, J. Shenker, S. and Turner, J. 2008. 'OpenFlow: Enabling Innovation in Campus Networks', in *SIGCOMM Computer Communication Review*, vol. 38/2, pp. 69–74.

Minsky, M. and Papert, S. 1969. *Perceptrons: An Introduction to Computational Geometry* (1st ed.), Cambridge, MA: The MIT Press.

Mishra, S. M., Sahai, A. and Brodersen, R. W. 2006. 'Cooperative Sensing among Cognitive Radios', in *IEEE International Conference on Communications (ICC'06)*, vol. 4, pp. 1658–1663.

Nakamoto, S. 2009. *Bitcoin: A Peer-to-Peer Electronic Cash System*. Available at: http://bitcoin.org/bitcoin.pdf Accessed at 19.11.2021.

OpenAI 2018. *AI and Compute*. Available at: https://blog.openai.com/ai-and-compute/ Accessed at 19.11.2021.

Roman, R., Zhou, J. and Lopez, J. 2013. 'On the Features and Challenges of Security and Privacy in Distributed Internet of Things', in *Computer Networks*, vol. 57/10, pp. 2266–2279.

Rosenblatt, F. 1958. 'The Perceptron: A Probabilistic Model for Information Storage and Organization in The Brain', in *Psychological Review*, vol. 65/5, pp. 386–408.

Rumelhart, D. E., Hinton, G. E. and Williams, R. J. 1986. 'Learning Representations by Back-Propagating Errors', in *Nature*, vol. 323, pp. 533–536.

Shi, W., Cao, J., Zhang, Q., Li, Y. and Xu, L. 2016. 'Edge Computing: Vision and Challenges', in *IEEE Internet of Things Journal*, vol. 3/5, pp. 637–646.

StorageCraft 2021. *7 Most Infamous Cloud Security Breaches*. Available at: https://blog.storagecraft.com/7-infamous-cloud-security-breaches/ Accessed at 19.11.2021.

Sweeney, L. 2002. 'K-Anonymity: A Model for Protecting Privacy', in *International Journal of Uncertainty, Fuzziness and Knowledge-Based Systems*, vol. 10/5, pp. 557–570.

TFDL 2018. *The Workshop on the Theoretical Foundation of Deep Learning*. Available at: http://pwp.gatech.edu/fdl-2018/ Accessed at 19.11.2021.

Turing, A. M. 1950. 'Computing Machinery and Intelligence', in *Mind*, vol. 59/236, pp. 433–460.

Vinyals, O., Toshev, A., Bengio, S. and Erhan, D. 2015. 'Show and Tell: A Neural Image Caption Generator', in *Computer Vision and Pattern Recognition 2015*. Computer Vision Foundation. Available at https://arxiv.org/abs/1411.4555. Accessed at 15.11.021.

Wei, J., Zhang, X., Ammons, G., Bala, V. and Ning, P. 2009. 'Managing Security of Virtual Machine Images in a Cloud Environment', in A. Oprea (ed.), *Proceedings of the 2009 ACM Workshop on Cloud Computing Security*, New York: ACM, pp. 91–96.

Weiser, M. 1991. 'The Computer for the 21st Century', in *Scientific American*, vol. 265/3, pp. 94–104.

Wood, G. 2021. *Ethereum: A Secure Decentralised Generalised Transaction Ledger*. Ethereum Project Yellow Paper. Available at: https://ethereum.github.io/yellowpaper/paper.pdf Accessed at 19.11.2021.

Yin, H., Jiang, Y. Lin, C., Luo, Y. and Liu, Y. 2014. 'Big Data: Transforming the Design Philosophy of Future Internet', in *IEEE Network*, vol. 28, pp. 14–19.

Zhuang, Y., Cappos, J., Rappaport, Th. S. and McGeer, R. 2013. *Future Internet Bandwidth Trends: An Investigation on Current and Future Disruptive Technologies*. Technical Report TR-CSE-2013-0411/01/2013, Department of Computer Sciences and Engineering, Polytechnic Institute of New York University, New York, NY.

5 Strategies for Managing Dynamic Complexity in Building the Internet

Ole Hanseth

5.1 Introduction

Increased processing power and higher transmission and storage capacity have made it possible to build increasingly integrated and versatile Information Technology (IT) solutions whose complexity has grown dramatically (BCS/RAE 2004; Hanseth and Ciborra 2007; Kallinikos 2007). Complexity can be defined here as the dramatic increase in the number and heterogeneity of included components, their relations and their dynamic and unexpected interactions in IT solutions. The Internet can be seen as a paradigm example of this growing complexity. Making the Internet evolve in favorable directions in the future is definitively a great challenge at the same time as the successful evolution of the Internet so far represents important lessons to be learned for how to guide the Internet's future, as well as how to cope with the growing complexity of IT solutions in general.

The growth in complexity has brought to researchers' attention novel mechanisms to cope with it, like architectures, modularity or standards (Baldwin and Clark 2000; Parnas 1972; Schmidt and Werle 1998). Another more recent stream of research has adopted a more holistic, socio-technical and evolutionary approach, putting the growth in the combined social and technical complexity at the center of an empirical scrutiny (see, for example, Edwards et al. 2007). These scholars view these complex systems as new types of IT artifacts and denote them with a generic label of *Information Infrastructures* (IIs). So far, empirical studies have garnered significant insights into the evolution of IIs of varying scale, functionality and scope, including Internet (Abbate 1999; Tuomi 2002), electronic marketplaces and Electronic Data Interchange (EDI) networks (Damsgaard and Lyytinen 2001; Wigand et al. 2006), wireless service infrastructures (Funk 2002; Yoo et al. 2005) or Enterprise Resource Packages (ERP) systems (Ciborra et al. 2000).

One challenge in the II research has been in the difficulty of translating vivid empirical descriptions of IIs evolution into effective *socio-technical* design principles that promote their evolution, growth and

DOI: 10.4324/9781003250470-7

complexity coordination. In this paper, we make some steps in addressing this challenge by formulating a new design approach to address the
dynamic complexity of IIs. From a technical viewpoint, designing an II
involves discovery, implementation, integration, control and coordination of increasingly heterogeneous IT capabilities. Socially, it requires
organizing and connecting heterogenous actors with diverging interests
in ways that allow for II growth and evolution. In the proposed
approach, we posit that the growing complexity of IIs originates from
local, persistent and limitless shaping of II's IT capabilities, due to the
enrollment of diverse communities with new learning and technical
opportunities. We argue that one common reason for the experienced II
design culprits is that designers cannot design IIs effectively by following traditional top-down design. In particular, the dynamic complexity
poses a chicken-egg problem for the would-be II designer that has been
largely ignored in the traditional approaches. On one hand, IT capabilities embedded in II gain their value by being used by a large number of
users demanding rapid growth in the user base (Shapiro and Varian
1999). Therefore, early on, II designers have to come up with solutions
that users may be persuaded to adopt while the user community is non-
existent or small. This requires II designers to address head on the needs
of the very first users before addressing completeness of their design, or
scalability. This can be difficult, however, because II designers must also
anticipate the completeness of their designs. This defines the *bootstrap
problem of II design*. On the other hand, when the II starts to expand
by benefitting from the network effects, it will switch to a period of
rapid growth. During this growth, designers need to heed for unforeseen
and diverse demands and produce designs that cope technically and
socially with these increasingly varying needs. This demands infrastructural flexibility, in that the II adapts technically and socially. This defines
the *adaptability problem of II design* (Edwards et al. 2007). Clearly,
these two demands contradict each other and generate tensions at any
point of time in II design (Edwards et al. 2007).

In this paper, we will address this tension by examining emergent
properties of IIs as *adaptive complex systems*. As IIs exhibit high levels
of dynamic complexity, they cannot be designed in the traditional way,
starting with a 'complete' set of requirements. II designers cannot
design IIs just based on the 'local' knowledge, but they can increase the
likelihood for successful emergence and growth of IIs by involving elements in their designs that take into account socio-technical *features* of
IIs generated by their dynamic complexity. We call this engagement
design for IIs.[1] While designing *for* IIs, the designers need to ask how
they can generate designs that promote continued growth and adaptation of IIs. To this end, we outline a set of design principles and rules
(Baldwin and Clark 2000; Markus et al. 2002). These principles and
rules can guide design behaviors in ways that allow IIs to grow and

adapt *as self-organizing systems*. We illustrate the validity of these design principles and rules by following the exegesis of Internet.

5.2 Design and Complex Adaptive Systems

We draw upon Complex Adaptive Systems (CAS) theory (Benbya and McKelvey 2006; Holland 1995). CAS addresses non-linear phenomena within physics and biology, but also in social domains, including financial markets (Arthur 1994). CAS investigates systems that *adapt* and *evolve* while they *self-organize*. The systems are made up of autonomous agents with the ability to adapt according to a set of rules in response to other agents' behaviors and changes in the environment (Holland 1995). Key characteristics of complex adaptive systems are: (1) non-linearity, i.e., small changes in the input or the initial state can lead to order of magnitude differences in the output or the final state; (2) order that emerges from complex interactions; (3) irreversibility of system states, i.e., that change is path dependent; and (4) unpredictability of system outcomes (Dooley 1996).

CAS helps characterize how the IIs can be initiated and how they grow and evolve while they self-organize. This is addressed by following the two principles: (1) create an attractor that feeds system growth to address the bootstrap problem; and: (2) assure that the emerging system will remain adaptable at 'the edge of chaos' while it grows to address the adaptability problem. We will next describe key categories of CAS theory and the logic that underpins these design principles and their 'because of' reasons, as suggested in Table 5.1.

5.2.1 Addressing Growth in II

A central claim in CAS is that order *emerges* – it is not designed by an omnipotent 'designer.' A typical example is the dynamic arrangements among cells and the establishment of standards without anyone ever intending to design them as such (e.g., QWERTY, TCP/IP). According to CAS, such orders emerge around *attractors,* i.e., a limited range of states within which the system growth can stabilize, and which allow the system to bootstrap (Holland 1995). De-facto standards (e.g., MS Windows, QWERTY, Internet standards) are examples of such attractors. Attractors stabilize a system through *feedback* loops (also called network effects or 'increasing returns'). In the case of standards, this happens because the value of a standard defining an IT capability depends on the number of users having adopted it. So, when a user adopts a standard, its value increases. This again makes it more likely that another user will adopt it, which further increases its value and so on (Arthur 1994; Shapiro and Varian 1999). A large installed base will also attract complementary IT

Table 5.1 CAS based design theory for dynamic complexity in II

Design goals	Bootstrap the IT capability into an installed base so that it gains momentum. Manage and allow for maximum II adaptability.
A set of system features	II as an unbounded, evolving, shared, heterogeneous and open recursively organized system of IT capabilities whose evolution is enabled and constrained by its installed base and the nature and content of its components and connections.
Kernel Theory of Flexible IIs	CAS informs how to address *bootstrap problem* in II designs by suggesting that: • The designer can gain momentum in the growth of II through attracting a critical mass of users. • The designer can enable non-linear growth by new combinations of the installed base. CAS informs how to address the *adaptability problem* in II designs by suggesting that. • The designer needs to recognize path dependencies within the installed base. • The designer needs to create lock-in through network externalities that exclude alternative pathways. • The designer needs to achieve modularity to accommodate the growing need for openness and heterogeneity in future.
Design principles	For the II bootstrap problem: 1. Design initially for usefulness 2. Draw upon existing installed base 3. Expand installed base by persuasive tactics For the II adaptability problem: 4. Make each IT capability simple 5. Modularize the II by building separately its principal functions and sub-infrastructures using layering, and gateways.

capabilities, thereby making the original capability increasingly attractive (Shapiro and Varian 1999). A larger installed base also increases the credibility associated with the capability and reduces user risks of foregone investments. Together, these features make an IT capability more attractive, leading to increased adoption that further increases its installed base (Grindley 1995). Some describe this process of getting 'the bandwagon moving'.

Positive network effects lead to self-reinforcing *path-dependent* processes. Overall, the involved path dependency suggests that past events – e.g., a serendipitous adoption, or correctly timed designs – can change history by generating irreversible effects, called *butterfly effects*. Such path-dependent growth will eventually lead to a *lock-in* when the adoption rates cross a certain threshold (David 1986). Such a lock-in happens when a system's growth reaches what Hughes (1987) calls a *momentum*. This creates a new lasting order with irreversible effects (Arthur 1994).

5.2.2 *Addressing Adaptability in II*

All systems evolve, but all systems do not adapt equally well. Systems that reach lock-in early on, or exhibit a large number of 'reverse salients' (Hughes 1987), will fail to do so. According to CAS, highly adaptable systems are characterized by the increased variety achieved through high modularity: 'variation is the raw material for adaptation' (Axelrod and Cohen 1999, p. 32). In other words, the larger the variety of agents and pathways for evolution, the more design alternatives can be tried out, and the more agents learn (Benbya and McKelvey 2006; Holland 1995). Accordingly, the larger the variety of IT capabilities and the larger the number of II designers, the larger is II adaptability. At the same time, a certain level of order is necessary in order to maintain stability in the design context. This stability is achieved through modularity. According to CAS, modularity creates a balance between variety and order by localizing the change and permitting fast and deep change in parts of the system.

5.3 Design Rules to Manage Dynamic Complexity – The Internet Case

The design principles listed in Table 5.2 guide II designers to conceive their designs in ways in which they can generate 'natural' order at the edge of chaos. To effectively carry out designs that conform to these principles, we need to break down the five design principles into *design rules* that govern designer's behaviors, influencing specific II components or their environments. In this section, we will articulate such design rules. Section 5.3.1 introduces some modularity concepts necessary in stating design rules. Section 5.3.2 offers a summary of their content and reports how we used Internet design to illustrate them. Section 5.3.3 introduces each design rule with illustrative examples from Internet design.

5.3.1 *Modularization of II*

In formulating the design rules, we need to distinguish properties of IIs that help modularize them. We, therefore, next define analytically types of (sub) IIs that – when composed *together* – allow to generate *modular* IIs. We recursively apply de-composition, i.e., identify separate subsets of IT capabilities within any II, which also are IIs, but which share a set of common functions and/or internal or external connections without having strong dependencies with the remaining IIs (Baldwin and Clark 2000; Kozlowski and Klein 2000).

Table 5.2 Design rules for dynamic complexity in the design for IIs

Design Problem	Element of CAS	Design Principles	Design Rules
Bootstrap problem Design goal: Generate attractors that bootstrap the installed base.	Create an IT capability that can become an attractor for the system growth.	1. Design initially for direct usefulness.	DR 1. Target IT capability to a small group. DR 2. Make IT capability directly useful without the installed base. DR 3. Make the IT capability simple to use and implement. DR 4. Design for one-to-many IT capabilities in contrast to all-to-all capabilities.
	Avoid dependency on other II components that deflect away from the existing attractors. Use installed base as to build additional attractors by increasing positive network externalities.	2. Build upon existing installed bases.	DR 5. Design first IT capabilities in ways that do not require designing and implementing new support infrastructures. DR 6. Deploy existing transport infrastructures. DR 7. Build gateways to existing service and application infrastructures. DR 8. Use bandwagons associated with other IIs.
	Exclude alternative attractors by persuasive tactics. Offer additional positive network externalities by expanding learning in the user community.	3. Expand installed base by persuasive tactics to gain momentum.	DR 9. 'Users before functionality' – grow the user base always before adding new functionality. DR 10. Enhance any IT capability within the II only when needed. DR 11. Build and align incentives so that users have real motivation to use the IT capabilities within the II in new ways. DR 12. Develop support communities and flexible governance strategies for feedback and learning.

Adaptability problem Design Goal: Make the system maximally adaptive and variety generating as to avoid technology traps.	Build capabilities that enable growth based on experience and learning. Use abstraction and gateways to separate II components by making them loosely coupled. Design IT capabilities and their combinations in ways that allow II growth. Use evolutionary strategies in the evolution of II that allow independent incremental change in separate components. Draw upon II designs that enable maximal variations at different components of the II.	4. Make the IT capability as simple as possible. 5. Modularize the II.	DR 13. Make the II as simple as possible in terms of its technical and social complexity by reducing connections and governance cost DR 14. Promote partly overlapping IT capabilities instead of all-inclusive ones. DR 15. Divide II recursively always into transportation, support and application infrastructures while designing the II DR 16. Use gateways between standard versions. DR 17. Use gateways between layers. DR 18. Build gateways between infrastructures. DR 19. Develop transition strategies in parallel with gateways.

We will first split IIs into *vertical application* IIs and *horizontal support* IIs. The former will deliver functional capabilities, which are deployable directly by one or more user communities. An example of application capability would be e-mail. The latter – support infrastructures – offer generic services, often defined in terms of protocols or interfaces necessary in delivering most, if not all, application services. They are primarily deployed by designer communities while building application capabilities[2] and include capabilities for data access and identification (addressing), transportation (moving) and presentation (formatting). They also constitute parts of the installed base that II designers need to take into account when bootstrapping an application infrastructure or making changes in support IIs.

We can further recursively decompose both application and support IIs. Thus, any II can be split into its application and support infrastructures until a set of 'atomic' IT capabilities are reached (per the recursive definition of II). In addition, any support infrastructure can be split into *transport* and *service* IIs. This split is justified, as transport infrastructure is necessary to make any service infrastructure work. The transport IIs offer data or message transportation services like the UDP/TCP/IP protocol stack (Leiner et al. 1997). On the other hand, service infrastructures support, for example, direct addressing, service identification, service property discovery, access and invocation, or security capabilities. They become useful when IIs start to grow in complexity and scale and designers need more powerful capabilities to configure application capabilities. A classic example of a service II is the Domain Name System (DNS) on the Internet, which maps mnemonic identifiers like Amazon.com to varying length bit representations, i.e., IP addresses. Both application and service IIs can be finally linked together horizontally through *gateways*. These offer flexible pathways for II expansion and navigation (Edwards et al. 2007; Hanseth 2001). An example of a gateway would be an IT capability, which supports multiple e-mail services running on different e-mail protocols.

5.3.2 Design Rules for Dynamic Complexity in II's

In total, we propose nineteen design rules for II dynamic complexity, shown in Table 5.2. Overall, the design rules characterize: (1) appropriate ways to organize and relate II components technically and socially (modular design and organized recursively) that address dynamic complexity similar to Baldwin and Clark's (2000) design rules; (2) desirable properties of specifications of II components (e.g., simplicity); (3) desirable sequences for design (e.g., design one-to-many IT capabilities before many-to-many IT capabilities); and (4) desirable ways to relate II specifications and associated components to one another (modularity and recursive application).

Below, we illustrate the deployment of each design rule by referring to episodes of Internet design. We chose Internet design as an illustration, as its design history offers rich insights into the situated application of the proposed design theory 'in use'. We chose to illustrate the theory with the design of Internet, because it qualifies as a grand 'success' story about II design *par excellence*. By any criterion, its design involved cultivating an II, which is shared, open, and heterogeneous, organized recursively and which operates without centralized control. We also content that the success of Internet testifies to the plausibility of the design rules discussed below.

5.3.3 Design Rules for Dynamic Complexity in the Design for II

We next discuss in detail how the nineteen design rules in Table 5.2 were inferred from the CAS theory offering a 'because of' justification for the design principles. To wit, each design rule offers a falsifiable statement of design outcomes to validate the theory. By analyzing whether the designer followed the rule and related design outcomes, we can determine whether the rule following did not lead to the predicted outcome, thus falsifying (partly) the proposed design theory.

5.3.3.1 Design Rules for the Bootstrap Problem

Often, new IT capabilities are not adopted despite their novelty, because users wait for others to adopt them first: Early adopters face high risks and costs but few benefits. In light of CAS, an II designer must generate attractors to propel users to adopt the IT capability so that its growth will reach a momentum (Hanseth and Aanestad 2003). We observe three design principles decomposed into twelve design rules that help generate and manage such attractors (see Table 5.2).

5.3.3.1.1 DESIGN RULES FOR PRINCIPLE 1: DESIGN INITIALLY FOR DIRECT USEFULNESS

Early users cannot be attracted to IT capability reasons like the size of their installed base. Therefore, we need design rules that foster relationships between the proposed IT capability and user adoption. Therefore, a small user population needs to be identified and targeted (Design Rule 1 (DR1[3])). The proposed IT capability has to offer the group *immediate* and *direct* benefits (DR2). Because first adopters accrue high adoption costs and confront high risks, the IT capability to be adopted must be simple, cheap and easy to learn (DR3). Here, cheap is defined in relation to both design and learning costs. Simple means that the design covers only the essential functionality expected and the capability is designed so that it is easy to integrate the IT capability with the installed base. Significant user investments cannot be expected because a small user base does not contribute either to the demand or the supply side economies of scale.

IT capabilities have varying impacts on the scale of increasing returns and the amount of positive feedback. They vary significantly between capabilities in which every user interacts symmetrically with every user (like e-mail) and capabilities in which one user interacts uni-directionally with the rest. Capabilities can also have multiple possible implementation sequences. In general, IT capabilities supporting asymmetrical interactions (one-to-many), and which are thus less dependent on network effects, should be implemented first, as the growth can be promoted locally (DR4). These capabilities have lower adoption barriers, as they do not need to reach a critical mass to generate fast adoption.

The Internet's success has been widely attributed to its successful bottom-up bootstrapping (Abbate 1999; Kahn 2006; Leiner et al. 1997; Tuomi 2002). Though Internet designers, early on, built bold scenarios of how the future of telecommunications would unfold (Tuomi 2002), the early uses of packet switching were targeted at small groups of researchers, who were interested in accessing powerful and expensive computers (DR1). The aim was to provide a limited range of directly useful IT capabilities: remote login and file transfer (DR1). Among these capabilities, remote login was a perfect choice, because each user could adopt it independently from one another and users had the skills and motivation to do so (DR3 and DR4). While the number of users grew, they could start to share data through file transfer. Later on, new capabilities were introduced in the same way (DR2). E-mail, for instance, was originally developed to support communications between persons responsible for maintaining the network when only four computers were connected to it (Abbate 1999) (DR1 and DR2). The design of transportation services (TCP) also followed an evolutionary approach, as multiple versions of increasingly complete protocols for TCP and IP were implemented in the early 1980s (DR3).

5.3.3.1.2 DESIGN RULES FOR PRINCIPLE 2: BUILD ON INSTALLED BASES

The second principle promotes connections with the existing installed base during design time. The II designer should thus design towards existing support infrastructures that the targeted user groups use (DR5). If an IT capability is designed so that it requires a new support infrastructure, this will erect heightened adoption barriers as, per our definition, the support infrastructure will be *sui generis* an II, which then needs to be bootstrapped with high learning barriers (Attewell 1992). As noted, transport infrastructures form the base for implementing II while the need for service infrastructures depends on the size and sophistication of application capabilities, or the size of the installed base. Whilst the installed base remains small, the II does not need advanced service infrastructures.

The II designer should therefore design towards the simplest possible service infrastructure (DR6). Next, capabilities associated with separate service and application infrastructures should be connected, when possible, through gateways, increasing connections between isolated user communities and benefitting adopters with larger positive network effects (DR7). As the designers link the new IT capabilities to the existing IIs, they need to take into account the speed and direction of the adoption of IT capabilities in neighboring infrastructures and capitalize on their bandwagon effects (DR8).

The Internet's early success resulted from exploiting established infrastructures as transport infrastructures (DR5) when TCP/IP was first implemented using modems over the telephone lines (Abbate 1999). In addition, each adopted capability has served to develop more advanced capabilities (Abbate 1999) (DR6). Currently, the Internet provides, for example, capabilities for electronic commerce, including transaction support (e.g., ebXML[4]), identification support (e.g., digital certificates), or security (e.g, SET), built as separate capabilities on top of TCP/IP and HTTP (Faraj et al. 2004; Nickerson and Zur Muehlen 2006). Another example is the initial growth of Internet's service infrastructures (DR6). In the beginning there were none and their need was discovered later when new service capabilities started to grow. Yet, the scale of Internet was still relatively small, so it was easy to design DNS capabilities and link it to a (now) stable transportation infrastructure. Later on, DNS became critical, as it increased flexibility of use through the management of dynamic IP addresses (using DHCP). The Internet designers have also increased the installed base through gateways (DR7). The expansion of the Web functionality is a case in point. The Web was originally thought to be useful for static information provisioning so that HTML tagged files could be downloaded using the HTTP protocol (Tuomi 2002). A significant added value for the Web was created by building gateways that leveraged upon data residing in organizational databases. This added dynamic or 'deep' web features: a call to database could be now embedded in HTML, as defined by Common Gateway Interface (CGI) specifications[5], and later expanded with Java standards (RMI[6]).

5.3.3.1.3 DESIGN RULES FOR PRINCIPLE 3: EXPAND INSTALLED BASE WITH PERSUASIVE ENROLLMENT TACTICS

After establishing the first attractor (usefulness), the II designers have to sustain growth. Therefore, when a simple version of the IT capability is available, the II designer needs to seek as many users as possible (DR9). This principle is captured well in the slogan 'users before functionality' emphasizing the criticality of generating positive network effects; the IT capability derives its value from the size of its user base, *not* from its superior functionality. New functionality should be added only when it is truly needed, and the original capability obtains new adoption levels so that the proposed capability will have enough users willing to cover the extra cost of design and learning (DR10). Many times, useful new functionality emerges when users start to deploy the IT capability in unexpected ways through learning by doing and trying, or through reorganizing the connections between the user communities and the IT capability (DR11). A growing installed base urges II designers to find means to align heterogeneous user interests and persuade them to continue to participate in the II. One approach is to use the installed base as a source of useful learning by creating user communities that offer feedback. This helps introduce new capabilities based on feedback and unexpected actor interactions (DR12) (Tuomi 2002; Zimmerman 2007).

Many capabilities during Internet design were established at times when the capabilities could be expected to work satisfactorily and serve a useful purpose (DR9). As a result, increasingly sophisticated application capabilities emerged, including Gopher (Minnesota), WAIS (Cambridge, Mass) and IRC (University of Oulu) (Rheingold 1993). As a result, Internet has grown enormously over the years in terms of new services and protocols. Typically, these capabilities emerged as local community responses to an identified local need (DR10), and only a tiny fraction of the Internet's current protocol stack was part of the initial specifications (DR10). The main reason for this was that most innovations took place at the 'edge', as design capability and application functionality were moved early on to the network boundary. New capabilities could be conceived and tried out whenever a user with a 'problem', and enough transportation capability, could leverage upon the new functionality (Tuomi 2002). Internet was also widely adopted by computer science and associated engineering communities as their research computing infrastructure (DR11 and DR12). The open packet switching standards turned out to be perfectly suited for the research vision shared by this movement (Kahn 2006).

5.3.3.2 Design Rules for the Adaptation Problem

When the bandwagon starts to roll, the II designers need to guarantee that the II will grow *adaptively* and reorganize constantly, with new connections between II components. Ad hoc designs, which were originally created for early users, will now threaten to create technology traps. If designers continue to generate highly interdependent and local IT capabilities, the whole system will become inflexible and reach a stasis. In contrast, if IT capabilities are organized modularly through loosely coupled 'layers', which can change independently, this will generate higher component variation for successful adaptation. The following two design principles decomposed into seven rules offer guidance to promote modularity.

5.3.3.2.1 DESIGN RULES FOR PRINCIPLE 4: MAKE THE ORGANIZATION OF IT
  CAPABILITIES SIMPLE

The first principle asks for the use of simple architectural principles during the initial design of the IT capabilities (DR13). It is easier to change
something that is simple than something that is complex. What makes a
collection of IT capabilities simple or complex is a function of its technical complexity, as defined by the number of its technical elements, their
connections and their rate of change (Edwards et al. 2007). Therefore,
following information hiding, simple interface protocols and functional
abstraction can help make the design simple. But it is just as important is
it to recognize the socio-technical complexity of the design space: the
number and type of connections between the technical and the social elements. In the lingo of Actor Network Theory (Latour 1999), the actor
network constituted by the II, i.e., its data elements, use practices, specifications and their discovery and enforcement practices, the relationships
to other infrastructures, the multiplicity of developers, the role of organizations, the variety of users, the regulatory bodies etc. – and a myriad of
links between all affect what can be changed and how (Latour 1999; Star
and Ruhleder 1996). Simpler actor networks can be created by making
them initially as small as possible and by keeping them loosely connected,
avoiding confrontations with competing networks. This is achieved by
pursuing separate specifications for distinct domains and separating the
concerns of different social and technical actors through functional
abstraction (Tilson 2008). Limiting the functional scope of application
infrastructures to a minimum keeps the related infrastructures separate.
Decomposing service IIs into a set of layers and separating their governance achieves the same goal. These principles decrease the technical
complexity of specifications but, more importantly, reduce their social
complexity. Finally, designs should promote partly-overlapping IT capabilities instead of all-inclusive ones. This increases variance and stimu-
lates innovation at different pockets by making it operate at 'the edge of
chaos' (DR14).

The principles of early Internet design promoted simplicity (DR13). Its
protocols were lean and simple and therefore had less ambiguity and
fewer errors. As a result, the implementations were simpler and were
easier to test and change. Origins of this approach date back to early
designs, which confronted early on the challenge of how to promote
change, but at the same time how to avoid technology traps. This was
expressed early in the Internet's specification approach:

> From its conception, the Internet has been, and is expected to remain,
> an evolving system whose participants regularly factor new require-
> ments and technology into its design and implementation.
>
> (RFC 1994, p. 6)

This vision was opposite to traditional design strategies in the telecommunication industry followed in the design of the ISO/OSI protocol stack, where designers assumed one homogeneous, complete and controllable network, which had to be completely specified (Abbate 1999; Russell 2006). This difference was later at the center of the controversy between the Internet community and the ISO/OSI committee (Russell 2006; Schmidt and Werle 1998). The OSI standardizers argued that Internet lacked critical functions; in contrast, the Internet community advocated technical simplicity and pragmatic value.

Many scholars have attributed the demise of the OSI to its disregard for this pragmatic approach (Rose 1992; Stefferud 1994). Finally, Internet always promoted designs that were partly overlapping, increasing variety. For example, it has generated several transportation protocols, e-mail protocols, information distribution protocols and so on (DR14).

5.3.3.2.2 DESIGN RULES FOR PRINCIPLE 5: MODULARIZE THE II

As noted, II designers need to organize capabilities into loosely coupled sub-infrastructures (Baldwin and Clark 2000; Parnas 1972). Therefore, IIs should be decomposed recursively into separate application, transport and service sub-infrastructures (DR15). Each II interface must hide mechanisms that implement these capabilities, so as to maintain loose couplings between the connected IIs. IIs need to be also decomposed vertically into independent neighboring application infrastructures, and II designers need to build gateways to connect them. Consequently, gateways must connect regions of II that run different versions of the same IT capabilities (DR16), or between different IT capability layers, e.g., transport or service (DR17), or between several dedicated application infrastructures (DR18) (Edwards et al. 2007). Finally, transitions between incompatible IT capabilities need to be supported by navigation strategies that allow local changes in different versions of the IT capability that run on the current installed base (DR19).

One reason for the speed of innovation in Internet was its initial modular design (DR15) (Tuomi 2002). The Internet's simple end-to-end architecture, which puts the 'intelligence' into the end nodes, has proven to be critical for its adaptive growth (DR15, DR16 and DR17) (Abbate 1999; David 2001). The design stimulated continued local application or service infrastructure innovation on top of the transport infrastructure (based on TCP/IP or UDP) (DR15) (Rheingold 1993; Tuomi 2002). Each of these capabilities was designed independently and its design decisions were insulated from potential changes in the underlying transport infrastructures. They were also governed separately.[7] The erection of the W3C and the governance of the web service community forms a case in point (Berners-Lee and Fischetti 1999). Gateways continue to play a critical role in the evolution of Internet; extensive use of gateways has prevented designers from acting like 'blind giants' and made early decisions easier to reverse (Hanseth 2001). Multiple gateways prevail, for instance, between the Internet's e-mail service and proprietary e-mail protocols (DR17). Another important family of gateways has been built between the Internet's access services and organization's applications and databases through web servers (DR18). Over the years, Internet protocols have been revised and extended (DR16 and DR19). One example is the revision of the transportation protocol from IPv4 to IPv6. The need to add new capabilities while attempting to overcome installed base inertia has been a major design challenge (Hovav and Schuff 2005; Monteiro 1998; RFC 1994). Between 1974 and 1978, four versions of the IP protocol were developed in fast experimental cycles until IPv4 was released (Kahn 1994). For the next 15 years, IPv4 remained stable. In the early 1990s, Internet's address space was expected to run out due to the Internet's exponential growth. Moreover, the addressing scheme in IPv4 did not support multicasting and mobility. This triggered a new round of designs to deliver a new IP version called IP version 6.[8] The final version, however, fulfilled only few of the original requirements – the most important one being the extension of the reverse salient – address space – to an awesome 2^{128} addresses.[9] The most important criterion in accepting the final specifications was in determining mechanisms that would introduce the new version in a stepwise manner (DR19), though initially this was not at all in the requirements (Hovav and Schuff 2005; RFC 1995; Steinberg 1995).[10]

5.4 Transformation of Large Scale IIs: Internet in Scandinavia

The growth of the Internet, in terms of users, services, nodes, providers, etc. (i.e., complexity) has made it increasingly harder to change and adapt to new requirements and to secure its future operations – it has become increasingly locked-in. In the Internet community, this process is referred to as the *ossification* of the Internet (see, for instance, Feldmann et al. 2009). It has become widely accepted in the Internet community that continued successful evolution of the Internet requires a different architecture than the existing one. What such an architecture should look like is definitively a hard question – how to transform the exiting Internet into a new and different one may be even harder one. As pointed out above, we believe that various forms of gateways, like the 'tunnelling' mechanism in the transformation from IPv4 to IPv6, will be critical elements in such a transformation process. Here, we will provide a rich illustration of the important roles gateways may play, by describing how a number of existing networks based on a variety of proprietary protocols were transformed into Internet (or ARPANET, as it was called at that time) in Scandinavia during the 1980s.

5.4.1 Background and Status, 1983–1985

In the late seventies and early eighties, most Nordic universities started to build computer networks. Different groups at the universities got involved in various international network building efforts. Around 1984, lots of fragmented solutions were in use and the level of use was growing. Obtaining interoperable services between the universities was emerging as desirable – and (technologically) possible.

The networks already in use – including their designers, users and operating personnel – were influential actors and stakeholders in the design and negotiations of future networks, including Nordunet. Here, we will briefly describe some.

5.4.1.1 Networks

IBM had set up and operated a network, which they called EARN. The network was based on IBM's proprietary technology (meaning RCSC protocols, which later on were redesigned and became widely known as the SNA protocol suite). It was connected to BITNET in the US. Most large European universities were connected. Users were found among many groups within the universities, collaborating with colleagues at other universities. The main services were e-mail and file transfer.

Another important network was HEPnet (High-Energy Physics Network). This was established to support collaboration among physicists, in particular among researchers related to the CERN lab outside Geneva, in Switzerland. The network was based on DECnet protocols.

EUnet was a network of Unix computers based on UUCP protocols. EUnet was mostly used by Unix users (doing software development), within academic institutions as well as private IT enterprises.

Norway was the first country outside the US linked to ARPANET (Abbate 1999; Spilling 1995). A node was set up in 1975 at the Norwegian Defence Research Establishment (NDRE) at Kjeller, outside Oslo. The second node was established by the department of informatics at the University of Oslo, connected to the node at Kjeller. Later on, more ARPANET nodes were set up. NDRE was using the net for research within computer communications, in collaboration with ARPA (in particular, communication via satellites). At that time, ARPANET was widely used among computer science researchers in the US, and computer science researchers in Norway very much wanted to get access to the same network to strengthen their ties to the US research communities. At that time, Unix was also diffusing rapidly. All Unix systems contained the ARPANET protocols, and most Unix computers were in fact communicating using these protocols in the local area networks they were connected to. Accordingly, there were lots of isolated IP islands in the Nordic countries. By linking these IP islands together, a huge Nordic network would be created.

In Norway, the development aiming to establish one network connecting all universities started in the early eighties. The goal was the establishment of one network linking every user and providing the same services to all. With this goal at hand, it was felt quite natural to link up with the ongoing effort aiming to develop the so-called Open Systems Interconnection (OSI) protocol standards within the framework of the International Standardization Organization (ISO) and then to build a network based on what would come out of that. An X.25 network was set up and an e-mail service was established in around 1984 or 1985 (based on the Canadian EAN system implementing the OSI X.400 standard).

5.4.1.2 Ideologies and Universal Solutions

As the networks described above were growing, the need for communication between users of different networks emerged. And the same was happening 'everywhere,' leading to a generally acknowledged need for one universal network providing the same universal services to everybody. Such a universal network required universal standards. So far, so good – everybody agreed on this. But what the universal standards should look like was quite a different issue.

This was a time of ideologies. The strongest ideology seems to be that of the ISO/OSI model, protocols and approach. In general, there was a religious atmosphere. Everybody agreed that proprietary protocols were bad and that 'open systems' were mandatory. The Americans pushed IP based technologies. They did so because they already had an extensive IP based network running and extensive experience from the design, operations and use of this network. The network worked very well (at least compared to others), and lots of application protocols were already developed and in use (FTP, Telnet, e-mail, etc.).

American research and university communities pushed IP, while both European researchers within the computer communications field and telecom operators pushed OSI. The role of telecom operators had the effect that the whole of the OSI protocol suite is based on 'telephone thinking.' The assumed importance of a connection-oriented transport protocol – as opposed to the connectionless mode of operation of TCP/IP – which will be mentioned in the next section, is one example illustrating this. (For more on this, see Abbate, 1995.) The Europeans wanted a non-IP based solution, believing that it would close the technological gap between Europe and the US.

5.4.2 Nordunet

The Nordunet initiative was taken by the top managers of the IT departments at the universities in the capitals of the Nordic countries. When the idea was accepted, funding was the next issue. The Ministry of the Nordic

Council was considered a proper funding organization. They had money – an application was written and funding was granted.

5.4.2.1 Strategy One: a Universal Solution, i.e., OSI

The Nordunet project was established in 1985. The people from the IT departments, having created the idea about the project, all believed in the OSI 'religion.' Next, they made an alliance with public authorities responsible for both the field that computer networks for research and education would fall into and for the funding institution (which was also closely linked to the authorities). Obtaining 'universal service' was an important objective for them; accordingly, they all supported the ideas behind OSI. This alliance easily agreed that an important element in the strategy was to unify all forces, i.e., enroll the computer communications researchers into the project. And so, this happened. As most of them were already involved in OSI related activities, they were already committed to the 'universal solution' objective and to following the OSI strategy to reach it.

However, products implementing OSI protocols were lacking. So, choice of strategy and, in particular, short term plans, were not at all obvious. To provide a proper basis for taking decisions, a number of studies looking at various alternative technologies for building a Nordic network were carried out:

* IP and other ARPANET protocols like SMTP (e-mail), FTP, and Telnet.
* Calibux protocols used in the JANET in the UK.
* EAN, an X.400 based e-mail system developed in Canada.

All these technologies were considered possible candidates for intermediate solutions only. The main rationale behind the studies was to find the best currently available technology. The most important criterion was the number of platforms (computers and operating systems) the protocols could run on.

Neither IP (and the other ARPANET) nor the Calibux protocols were found to be acceptable. The arguments against IP and ARPANET were, in general, that the technology had all too limited functionality. FTP had limited functionality compared to OSI's FTAM protocol (and also compared to the Calibux file transfer protocol, which FTAM's design, to a large extent, was based on). The Nordunet project group, in line with the rest of the OSI community, found the IP alternative 'ridiculous,' considering the technology to be all too simple and to not be offering the required services. There were, in particular, hard discussions about whether the transport level services should be based on connection-oriented or connectionless services. The OSI camp argued that connection-oriented services were the most important. IP is based on a connectionless data-gram service, which the IP camp considered one of the strengths of the ARPANET technology.

The main argument against Calibux was that the protocols did not run on all required platforms (computers and operating systems). One important constraint put on the Nordunet project was that the solutions should be developed in close cooperation with similar European activities. This made it almost impossible to go for ARPANET protocols.

The IP camp believed that IP (and the other ARPANET protocols) was the universal solution needed, and that the success of ARPANET proved this.

The users were not directly involved in the project, but their views were important in making the project legitimate. They were mostly concerned about services. They wanted better services – right away! But in line with this they also argued that more efforts should be put into the extensions and improvements of the networks and services they were using already and less into the long-term objectives. The HEPnet users expressed this most clearly. They were using DECnet protocols and DEC computers (in particular, VAX). DEC computers were popular among most Nordic universities and, accordingly, they argued that a larger DECnet could easily be established and that this would be very useful for large groups. The physicists argued for a DEC solution. Nobody argued, however, for a 'clean' DECnet solution as a long-term objective.

Both on the Nordic and the global scene (Abbate 1995), the main fight was between the IP and OSI camps. This fight involved several elements and reached far beyond technical considerations related to computer communications. At all universities, there was a fight and deep mistrust between IT departments and computer science departments. The IT departments were concerned about delivering ('universal') services to the whole university as efficiently as possible. They thought this could be done best by one coherent set of shared services, delivered by a centralized organization. Most of the time, the computer science departments found the services provided by the IT departments to be lagging behind the technological edge and unsatisfactory in relation to their requirements. They saw themselves as rather different from the other departments, as computers were their subject. They had different requirements and competencies, believing that they would be much better off if they were allowed to run their own computers. But the IT departments were very afraid of losing control if there were any computers outside the domain they ruled.

The computer science departments also disagreed with the IT departments about what should be in focus when building communication services and networks. The IT departments focused first on their own territory, then on the neighboring area. This meant first establishing networks across the university and then, secondly, extending and enhancing this so that it became linked to the networks at the other universities in Norway, then the Nordic countries and so on. The computer science departments, however, were not primarily interested in communicating with other departments at the same university. They wanted first of all to communicate and collaborate with fellow researchers at other computer science departments – not primarily in Norway or other Nordic countries either, but in the US. They wanted Unix computers to run the same software as their colleagues in the US, and they wanted the connection to ARPANET to communicate with them.

The IT departments would not support Unix as long as it was not considered feasible as the single, 'universal' operating system for the whole university, and they would not support IP for the same reason. Thirdly, they wanted complete control and would not let the computer science department do it by their own either. To get money to buy their own computers, the computer science department had to hide this in applications for funding of research projects within VLSI and other fields. The fight over OSI (X.25) and IP was deeply embedded into these networks of people, institutions and technologies.

As all intermediate solutions were dismissed, it was decided to go directly for an OSI based solution. The first version of the network would be build based on X.25 and the EAN system providing e-mail services. This solution was very expensive, and the project leaders soon realized that it did not scale. X.25 was full of trouble. The problems were mostly related to the fact that the X.25 protocol specification is quite extensive and, accordingly, lead easily to incompatible implementations. Computers from several vendors were used within the Nordunet community, and there were several incompatibilities among the vendors' implementations. Further trouble was caused by the fact that lots of parameters have to be assigned values when installing/configuring an X.25 protocol installation. To make the protocol installations interoperate smoothly, the parameter setting has to be coordinated. In fact, they required coordination beyond the capabilities of the Nordunet project.

5.4.2.2 *Strategy Two: Intermediate, Short-term Solutions – The Nordunet Plug*

The project worked on the implementation of the network as it was specified for about a year or so without any significant progress. The standardization of OSI protocols was also (continually) discovered to be more difficult and the progress slower than expected, making the long-term objectives continually more distant. The Nordic Council of Ministers was seriously discussing stopping the project because of the lack of progress. New approaches were desperately needed.

At the same time, other things happened. IBM wanted to transfer the operations of its EARN network to the universities. The project managers had, over some time, developed the idea to use EARN as backbone of a multi-protocol network. They started to realize that OSI would take a long time – one *had to* provide services before that. OSI was, all the time, ideological important, but one had to become more and more pragmatic. The idea about 'The Nordunet Plug' was developed. This idea means that there should be one 'plug' common for everybody that would hook onto the Nordunet network. The plug should have 4 'pins,' one for each of the network protocols to be supported: OSI/X.25, EARN, DECnet and ARPANET/IP.

The idea was presented as if the plug implemented a gateway between all the networks. That was, however, not exactly the case. The plug only provided access to a shared backbone network. An IBM computer running EARN/RSCS protocols could communicate only with another computer also running the same protocols. There was no gateway enabling communications between, say, an RSCS and an IP based network.

The EARN idea received strong support. The project got hold of the EARN lines through some very quick decisions, and the implementation of a Nordic network based on the 'Nordunet plug' idea started. They succeeded in finding products that made its implementation quite straight forward. First, Vitalink Ethernet bridges were connected to the EARN lines. This means that Nordunet was essentially an Ethernet. To these Vitalink boxes the project linked IP routers, X.25 switches and EARN 'routers.' For all these protocols, there were high quality products available that could be linked to the Vitalink Ethernet bridges.

This solution had implications beyond enabling communication across the backbone. Further, the EARN backbone included a connection to the rest of the global EARN network. A shared Nordic line to ARPANET was established and connected to the central EARN node in Stockholm. 64 Kb lines to CERN for HEPnet were also connected.

Having established a shared backbone, the important next step was, of course, the establishment of higher-level services like e-mail, file transfer, remote job entry (considered very, very important at that time for sharing computing resources for number crunching), etc. As most of the networks in use had such services based on proprietary protocols, the task for the Nordunet project was to establish gateways between these. A large activity was set up, aiming to do exactly that. When gateways at the application level were established, full interoperability would be achieved. A gateway at the transport level would do the job if there were products available at the application level (e-mail, file transfer, etc.) on all platforms, implementing the same protocols. Such products did not exist.

Before this, users in the Nordic countries used gateways running on computers in the US to transfer e-mail between computers running different e-mail systems. That meant that when sending an e-mail between two computers standing next to each other, the e-mail had to be transferred across the Atlantic, converted by the gateway in the US and finally transferred back again. The Nordunet established a gateway service between the major e-mail systems used. The service was based on software developed at CERN.

File transfer gateways are difficult to develop, as they require conversion on the fly. CERN had a gateway between file transfer protocols, called GIFT (General Interface for File Transfer), running on VAX/VMS computers. An operational service was established at CERN. It linked the services developed within the Calibux network (often called 'Blue book'), DECnet, the Internet (FTP), and EARN networks. The gateway worked very well at CERN. Within Nordunet, the Finnish partners were delegated the task of establishing an operational gateway service based on the same software. This effort was, however, given up, as the negotiations about conditions for getting access to the software failed. In spite of this, a close collaboration emerged between the Nordunet project and CERN people. They were 'friends in spirit' – having OSI as the primary long-term objective, but at the same time concentrating on delivering operational services to the users.

5.4.2.3 *From an Intermediate to a Permanent Solution*

When the Nordunet Plug was in operation, a new situation was created. Users started to use the network. The network services had to be maintained and operated. And users' experiences and interests had to be accounted for when making decisions about the future changes to the network. The maintenance and operation work, as well as the use of the network, was influenced by the way the network – and, in particular, the 'plug' as its core – were designed. The 'plug' became an actor playing a central role in the future of the network.

Most design activities were directed towards minor, but important and necessary, improvements of the net that its use disclosed. Fewer resources were left for working on long-term issues. However, in the Nordunet community, long-term issues were still considered important, and the researchers involved continued their work on OSI protocols and their standardization. The war between Internet/IP and OSI/X.25 continued. The OSI supporters believed as strongly as ever that OSI, including X.25, was the ultimate solution. Some, however, turned their focus more towards practical solutions, in particular making bridges to the growing IP communities.

In parallel with the implementation and early use phase of the 'plug,' other things happened. Unix diffused fast in academic institutions, and the ARPANET was growing fast as its protocols were implemented on more platforms and created more local IP communities (in LANs). On the other hand, there was, in practical terms, no progress within the OSI project.

The increased availability of IP on more platforms led to an increase in use of 'dual stack' solutions, i.e., installing more than one protocol stack on a computer, linking it to more than one network. Each protocol stack is then used to communicate with specific communities. This phenomenon was common, in particular, among users of DEC computers. Initially, they were using DECnet protocols to communicate with locals or, for instance, fellow researchers using HEPnet and using IP to communicate with ARPANET users.

The shared backbone, the e-mail gateway and 'dual stack' solutions created a high degree of interoperability among Nordunet users. Individual users could, for most purposes, choose which protocols they preferred – they could switch from one to another based on personal preferences – and as IP and ARPANET were diffusing fast, more and more users found it most convenient to use IP. This led to a smooth, unplanned and uncoordinated transition of the Nordunet into an IP based network.

One important element behind the rapid growth of the use of IP inside Nordunet was the fact that ARPANET's DNS service made it easy to scale up an IP network. In fact, this can be done by just giving a new computer and address, hooking it on and entering its address and connection point into DNS. No change is required in the rest of the network. All the network needs to know about the existence of the new node is taken care of by DNS. For this reason, the IP network could grow without requiring any work by the network operators, and the OSI enthusiasts could not do anything to stop it either.

5.4.3 Lessons to Be Learned

The Nordic countries have been among those where computer networks in general and the Internet in particular have been most widely diffused and most heavily used. (In some periods there have been relatively more Internet users in Norway and Sweden than even in the US.) I see the Nordunet project and the 'plug' in particular as a most important explanation of this. I will, in this section, discuss in general terms the important role the 'plug' played as a gateway, but I will also discuss gateways beyond what is illustrated by the 'plug' and the Nordunet project.

The most important role the 'Nordunet Plug' was playing as a gateway in this case was the way it enabled and supported a 'rational' process for designing these kinds of large-scale infrastructures.

5.4.3.1 Allowing Experimentation and Learning

Experimental design of large-scale networks and infrastructures is, indeed, problematic but nevertheless the most important success criterion. It is absolutely impossible to develop several different alternatives and test them. To find out what works well in the real world, the solutions have to be used in real work by, perhaps, thousands of users. The development of the kind of technological solutions the Nordunet aimed for was about designing something with a significant degree of novelty. Accordingly, a working solution could not be developed without acquiring the knowledge that was lacking. What is a good and effective solution is an empirical question – it cannot be derived from theory, so some kind of experimentation is the only way to produce this knowledge. This dilemma has to be solved. On a general level, this can be done by adopting some kind of evolutionary approach. This means that we first develop one piece, and this piece is at the level of complexity where experimentation is not too costly. The problem with this approach is that it may lead us into path-dependent processes and lock-ins. Each piece developed has to be integrated with the existing ones and, accordingly, design assumptions made when developing the early pieces may turn out to be inappropriate for the later ones. But at that time, the investments made are too big to change this.

The Nordunet project team adopted a mere specification driven approach – first agree on needs, then specify the solutions, then implement and use them. Other actors involved advocated more evolutionary design models, arguing that the overall solution should be based on the one they were already using. But this strategy did not work, because those already using a network could not accept any of the others – the costs they had to pay to switch were too high. Accordingly, another kind of evolutionary process had to be adopted and the fact that the Nordunet Plug enabled this is the main rationale behind its success. First, the Plug gave all users access to a shared backbone. This was an important achievement from the perspective of all actors being involved because:

- The solution could easily be implemented – the knowledge and technology needed were there.
- It offered the users improved services. all of them could through the Plug and the backbone reach new communicating partners.
- It was a step forward towards the 'final' solution independent of how this would look. From the solution established one could – in principle – just as easy move towards any of the alternatives advocated. For this reason, the solution was a compromise that all actors involved could easily accept.
- It allowed extended use of all existing alternatives, so it enabled extended experimentation and generation of new knowledge about how the 'final' solutions should look – or at least what the most appropriate next steps could be.

I mentioned above that evolutionary development processes may lead to a state of lock-in. Gateways are also important tools in this respect because they help to transform one network form running one protocol standard into another network running another standard.

5.4.3.2 *User Involvement and Democratic Design Processes*

Large-scale information infrastructures have many users. User influence and involvement in technological design is a topic involving many issues. One aspect is the fact that it is only the users who can decide whether or not a specific technology is appropriate for their work tasks (or other kinds of use processes). Accordingly, they must be involved in the design process in one way or another to inform the designers about their needs and to judge whether a specific technology is acceptable or not. In addition, how and to what extent users have been involved in the design of a solution has significant implications on how the organizational implementation process unfolds, how the solution is used and what benefits are gained. The general trend is that the more the users have been involved, the more positive they are towards the use of the technology, the smoother the implementation process is running and the more active the users are in learning how to use the solution in the most beneficial ways. User participation in – and influence over – the design process may also be argued from a political perspective. Such participation is simply a democratic right.

User influence is, as argued in the previous section, an illusion, unless it is based on practical use of one or more alternative solutions in a realistic setting. Users can only give designers trustworthy advice if they have practical experience with the technology.

All these principles apply to the design of infrastructures just as much as any other technology. Compared to information systems, it might be seen as even more important. Within an organization, management may decide that a specific solution should be developed and/or implemented and instruct the departments to use it. Concerning the design and implementation of an infrastructure, those involved in projects like Nordunet, or the OSI standardization, have a very weak influence over potential users. If the users do not like the technology, they will just not use it.

For all these reasons, an evolutionary process, which gateways are enabling, is crucial.

5.4.3.3 *Adaptation to Changing Environments*

All technologies are designed to satisfy needs derived from specific ways of living or working. Over time, these conditions change and technologies need to adapt to them to remain useful and appropriate. So too is the case with infrastructures. This implies that standardized network protocols also need to be changed. The ongoing transformation of the Internet from running IP version 4 to the new IP version 6 is a paradigm example of this (Hanseth et al. 1996; Monteiro 1998). The new IP protocol is required primarily because the growth of the Internet requires an extended address space. But support beyond what the old version of IP is offering is required by several emerging services and ways of using the Internet. This includes (enhanced) support for security, accounting, broadband networks, real time multi-media, etc. (and not all of these are supported satisfactorily by the new version either.)

The IP example also illustrates the fact that an infrastructure generates new requirements as its range and use grow.

As one network or infrastructure and its use grow, the infrastructure will also 'meet' other networks that were initially assumed to be independent and generate needs for integrating these. This may be seen as the history of the different networks that were in use before the Nordunet project started and that the user communities wanted to integrate (i.e., HEPnet, EARN, etc.).

Gateways are key tools to enable the change of an infrastructure and (some of) its standards from one version to another and, accordingly, avoid being trapped in a lock-in situation.

5.4.3.4 *Living with Heterogeneity: Gateways as Final Solutions*

Gateways are important beyond representing compromises and being tools enabling transitions towards 'final' solutions; they are often also important components of such 'final' solutions. Gateways are interfaces between different, but related, services. The world is heterogeneous. There is a high number of computer networks and information infrastructures around – and the number is currently rapidly increasing. Many of the networks are, and will be, related yet still different. This means that they should be connected to enable some kind of interoperation at the same time as they should be kept as separate as possible to make their co-evolution with their changing environment easier and simpler. This is what happened with the development of AC and DC together with the rotary converter. Another example is EDI networks in healthcare. Most information in such networks is exchanged between institutions within the same country. However, some information is also exchanged across national borders. So, in principle, one might argue that there should be only one set of coherent and consistent global standards. A definition of such a coherent set of standards has indeed been attempted at the European level – with modest success, to say the least (Hanseth and Monteiro 1997). The problem is that the variation among national healthcare systems requires very complex standards if the standards are to satisfy all needs. Experience so far indicates that the standards become all too complex to be implemented at all (ibid.). An alternative approach would be to develop national standards and then gateways to enable the (limited – but slowly increasing) transnational information exchange.

When larger networks are already built, linking them through gateways may be easier and cheaper than moving to one shared protocol, although the functionality will be well covered by one such universal standard. This seems to be the case in the world of e-mail. Several protocols and networks exist: the Internet based on SMTP, X.400 based networks, and networks based on proprietary protocols and systems like cc:mail Notes, Microsoft Exchange, etc. Gateways are established between these networks and protocols. Although they are not translating perfectly, most users, in practice, experience their e-mail system as providing access to a service through which any e-mail user can be reached.

Another example may be the 'global' web. This network gives us access to any information – in any database – through a web browser. A key tool enabling this is various gateways between web servers and the databases where the information is actually stored. Gateway technologies here include, in particular, CGI scripts. These gateways link together networks internally in an organization, as well as between organizations. One might imagine that there could be one universal solution used to access a database both from outside and inside an organization. However, this is not a good solution for several reasons:

The Web technology is developing rapidly. How to integrate corporate networks and the global Internet is still an unsettled issue. How to do this is a matter of experiment over a long period of time. Such experiments require modular and flexible solutions – one must be able to change the modules independently. This requires interfaces between the modules, and what is going on inside the interface is irrelevant for outsiders. Gateways are the exact interfaces needed to make larger networks flexible.

There is no reason to believe that the requirements from and service provided to insiders and outsiders will be basically the same. Accordingly, insiders and outsiders will see a system as completely different and will therefore need to get access to it through different technologies.

Notes

1 In one sense, infrastructures just evolve, if we rely on biological metaphor and the idea of 'blind' mutation. But, because infrastructures are artifacts created by intentional action, we prefer to use the term 'design for' instead of 'design of' as designer's behaviors matter how, and to what extent, the infrastructure can evolve. We have elsewhere proposed the term 'cultivate' for this type of design activity.
2 This principle is similar to decomposition of dedicated IT applications if we distinguish between user defined computational functions (e.g., computing a salary) and generic horizontal system functions (e.g., retrieving or storing an employee record).
3 This refers to the rule label in Table 5.3.
4 See http://www.ebxml.org/ Accessed at 3 August 2021.

5 See http://hoohoo.ncsa.uiuc.edu/cgi/overview.html Accessed at 3 August 2021.
6 See http://java.sun.com/products/jdk/rmi/ Accessed at 3 August 2021.
7 This was not always done without friction (see, for example, Nickerson and Zur Muehlen 2006).
8 For a definition, see http://playground.sun.com/pub/ipng/html/ipng-main.html Accessed at 3 August 2021.
9 It has been later observed that other 'workarounds' like DHCP and NAT actually could circumvent the address space problem and the value of IPv6 in this solving the original requirement has been questionable.
10 See, for example, http://www.ipv6.org/ Accessed at 3 August 2021.

References

Aanestad, M., and Hanseth, O. 2002. 'Growing Networks: Detours, Stunts and Spillovers', in *Proceedings of COOP 2002, Fifth International Conference on the Design of Cooperative Systems*, St. Raphael, France, 4–6 June 2002.

Abbate, J. 1995. 'The Internet Challenge: Conflict and Compromise in Computer Networking', in Summerton, J. (ed.), *Changing Large Technical Systems*. Oxford: Westview Press, pp. 193–210.

Abbate, J. 1999. *Inventing the Internet*. Cambridge, MA: The MIT Press.

Arthur, W. B. 1994. *Increasing Returns and Path Dependence in the Economy*. Ann Arbor, MI: The University of Michigan Press.

Attewell, P. 1992. 'Technology Diffusion and Organizational Learning: The Case of Business Computing', in *Organization Science*, vol. 3/1, pp 1–19.

Axelrod, R. M., and Cohen, M. D. 1999. *Harnessing Complexity: Organizational Implications of a Scientific Frontier*. New York: Free Press.

Baldwin, C., and Clark, K. 2000. *Design Rules*. Cambridge, MA: The MIT Press.

BCS/RAE. 2004. 'The Challenges of Complex IT Projects', in *British Computer Society and Royal Academy Engineering Project*. Available at http://www.bcs.org/upload/pdf/complexity.pdf Accessed at August 2009.

Benbya, H., and McKelvey, B. 2006. 'Toward Complexity Theory of Information System Development', in *Information, Technology and People*, vol. 19/1, pp. 12–34.

Berners-Lee, T. with Fischetti, M. 1999. *Weaving the Web: The Original Design and Ultimate Destiny of World Wide Web by Its Inventor*. San Francisco, CA: Harper-Collins.

Ciborra, C., Braa, K., Cordella, A., Dahlbom, B., Failla, A., Hanseth, O., Hepsø, V., Ljungberg, J., Monteiro, E., and Simon, K. 2000. *From Control to Drift. The Dynamics of Corporate Information Infrastructures*. Oxford: Oxford University Press.

Damsgaard, J., and Lyytinen, K. 2001. 'Building Electronic Trading Infrastructure: A Private or Public Responsibility', in *Journal of Organizational Computing and Electronic Commerce*, vol. 11/2, pp. 131–151.

David, P. A. 1986. 'Understanding the Economics of QWERTY', in Parker, W. N. (ed.), *Economic History and the Modern Economist*. Oxford: Basil Blackwell.

David, P. A. 2001. The Beginnings and Prospective Ending of "End-to-End". Working Papers 01012, Stanford University, Department of Economics. Available at http://ideas.repec.org/p/wop/stanec/01012.html Accessed at 3.8.2021.

Dooley, K. 1996. 'Complex Adaptive Systems: A Nominal Definition', in *The Chaos Network*, vol. 8/1, pp. 2–3. Available at http://www.public.asu.edu/~kdooley/papers/casdef.PDF. Accessed at 13.8.2009.

Edwards, P., Jackson, S., Bowker, G., and Knobel, C. 2007. 'Understanding Infrastructure. Dynamics, Tension and Design. Report of a Workshop on "History and theory of infrastructures: Lessons for new scientific infrastructures"', University of Michigan, School of Information, January 2007. Available at http://www.si.umich.edu/InfrastructureWorkshop/documents/Understanding-Infrastructure2007.pdf Accessed at 15.3.2007.

Evans, D. S., Hagiu, A., and Schmalensee, R. 2006. *Invisible Engines: How Software Platforms Drive Innovation and Transform Industries*. Cambridge, MA: The MIT Press.

Faraj, S., Kwon, D., and Watts, S. 2004. 'Contested Artifact: Technology Sensemaking, Actor Networks, and the Shaping of the Web Browser', in *Information Technology & People*, vol. 17/2, pp. 186–209.

Feldmann, A., Kind, M., Maennel, O., Schaffrath, G., and Werle, Ch. 2009. 'Network Virtualization – An Enabler for Overcoming Ossification', in *ERCIM News*, vol. 77, pp. 21–22.

Funk, J. L. 2002. *Global Competition between and within Standards: The Case of Mobile Phones*. New York: Palgrave.

Grindley, P. 1995. *Standards, Strategy, and Politics. Cases and Stories*. New York: Oxford University Press.

Hanseth, O. 2000. 'The Economics of Standards', in Ciborra, C., Braa, K., Cordella, A., Dahlbom, B., Failla, A., Hanseth, O., Hepsø, V., Ljungberg, J., Monteiro, E., and Simon, K. (eds.), *From Control to Drift. The Dynamics of Corporate Information Infrastructures*. Oxford: Oxford University Press, pp. 56–70.

Hanseth, O. 2001. 'Gateways – Just as Important as Standards. How the Internet Won the "Religious War" about Standards in Scandinavia', in *Knowledge, Technology and Policy*, vol. 14/3, pp. 71–89.

Hanseth, O., and Aanestad, M. 2003. 'Bootstrapping Networks, Infrastructures and Communities', in *Methods of Information in Medicine*, vol. 42, pp. 384–391.

Hanseth, O., and Ciborra, C. (eds.), 2007. *Risk, Complexity and ICT*. Northampton, MA: Edward Elgar.

Hanseth, O., and Lundberg, N. 2001. 'Information Infrastructure in Use – An Empirical Study at a Radiology Department?', in *Computer Supported Cooperative Work (CSCW). The Journal of Collaborative Computing*, vol. 10/3–4, pp. 347–372.

Hanseth, O., and Monteiro, E. 1997. 'Inscribing Behaviour in Information Infrastructure Standards', in *Accounting Management and Information Technology*, vol. 7/4, pp. 183–211.

Hanseth, O., Monteiro, E., and Hatling, M. 1996. 'Developing Information Infrastructure: The Tension between Standardization and Flexibility', in *Science, Technology and Human Values*, vol. 21/4, pp. 407–426.

Holland, J. 1995. *Hidden Order*. Reading, MA: Addison-Wesley.

Hovav A., and Schuff D. 2005. 'The Changing Dynamic of the Internet: Early and Late Adopters of Ipv6 Standard', in *Communications of the Association of the Information Systems*, vol. 15, art. 14, pp. 242–262.

132   *Ole Hanseth*

Hughes, T. P. 1987. 'The Evolution of Large Technical Systems', in Bijker, W. E., Hughes, T. P., and Pinch, T. (eds.), *The Social Construction of Technological Systems*. Cambridge, MA: The MIT Press, pp. 51–82.

Kahn, R. E. 1994. 'The Role of Government in the Evolution of the Internet', in *Communications of the ACM*, vol. 37/8, pp. 415–419. Special issue on Internet technology.

Kallinikos, J. 2007. 'Technology, Contingency and Risk: The Vagaries of Large-Scale Information Systems', in Hanseth, O., and Ciborra, C. (eds.), *Risk, Complexity and ICT*. Northampton, MA: Edward Elgar, pp. 46–74.

Kayworth, T., and Sambamurthy, S. 2000. 'Facilitating Localized Exploitation of Enterprise Wide Integration in the Use of IT Infrastructures: The Role of PC/LAN Infrastructure Standards', in *The Data Base for Advances in Information Systems*, vol. 31/4, pp. 54–80.

Kozlowski, S., and Klein, K. (2000). 'A Multi-Level Approach to Theory and Research in Organizations: Contextual, Temporal and Emergent Processes', in Klein, K., and Kozlowski, S. (eds.), *Multilevel Theory, Research and Methods in Organizations*. San Francisco: Jossey-Bass, pp. 3–90

Latour, B. 1999. *Pandora's Hope. Essays on the Reality of Science Studies*. Cambridge, MA: Harvard University Press.

Leiner, B. M., Cerf, V. C., Clark, D. D., Kahn, R. E., Kleinrock, L., Lynch, D. C., Postel, J., Roberts, L. G., and Wolff, S. S. 1997. 'The Past and Future History of the Internet', in *Communications of ACM*, vol. 40/2, pp. 102–108.

Monteiro, E. 1998. 'Scaling Information Infrastructure: The Case of the Next Generation IP in Internet', in *The Information Society*, vol. 14/3, pp. 229–245.

Nickerson, J. V., and zur Muehlen, M. 2006. 'The Social Ecology of Standards Processes: Insights from Internet Standard Making', in *MIS Quarterly*, vol. 30/5, pp. 467–488.

Parnas, D. L. 1972. 'A Technique for Software Module Specification with Examples', in *Communications of the ACM*, vol. 15/5, pp. 330–336.

RFC. 1994. *The Internet Standards Process – Revision 2*. RFC 1602, IAB and IESG.

RFC. 1995. *The Recommendation for the IP Next Generation Protocol*. RFC 1752, IAB and IESG.

Rheingold, H. 1993. *The Virtual Community: Homesteading the Electronic Frontier*. Reading, MA: Addison Wesley.

Rose, M. T. 1992. 'The Future of OSI: A Modest Prediction', in *Proceedings of the USENIX Conference 1992*, USENIX Association, Berkley, 1 June 1992.

Russell, A. 2006. '"Rough Consensus and Running Code" and the Internet – OSI standards War', in *IEEE Annals if the History of Computing*, July–September, pp. 48–61.

Schmidt, S. K., and Werle, R. 1998. *Coordinating Technology. Studies in the International Standardization of Telecommunications*. Cambridge, MA: The MIT Press.

Shapiro, C., and Varian, H. R. 1999. *Information Rules: A Strategic Guide to the Network Economy*. Boston, MA: Harvard Business School Press.

Spilling, P. 1995. *Fra ARPANET til internett, en utvikling sett med norske øyne*, Note, Telenor forskning, 25 mars. Available at http://www.isoc-no.no/isoc-no/social/arpa-no.html

Star, L. S., and Ruhleder, K. 1996. 'Steps toward an Ecology of Infrastructure: Design and Access of Large Information Spaces', in *Information Systems Research*, vol. 7/1, pp. 111–134.

Stefferud, E. 1994. 'Paradigms Lost', in *Connexions. The Interoperability Report*, vol. 8/1.

Steinberg, S. G. 1995. 'Addressing the Future of the Net', in *WIRED*, vol. 3/5, pp. 141–144.

Tilson, D. 2008. *Reconfiguring to Innovate: Innovation Networks during the Evolution of Wireless Services in the United States and the United Kingdom.* PhD Thesis, Department of Information Systems, Case Western Reserve University.

Tuomi, I. 2002. *Networks of Innovation. Change and Meaning in the Age of the Internet.* Oxford: Oxford University Press.

Wigand, R. T., Lynne Markus, M., Steinfield, C. W., and Minton, G.. 2006. 'Standards, Collective Action and IS Development-Vertical Information Systems Standards in the US Home Mortgage Industry', in *MIS Quarterly*, Special Issue on Standards and Standardization, vol. 30, pp. 439–465.

Yoo, Y., Lyytinen, K., and Yang, E. 2005. 'The Role of Standards in Innovation and Diffusion of Broadband Mobile Services: The Case of South Korea', in *Journal of Strategic Information Systems*, vol. 14/2, pp. 323–353.

Zimmerman, A. 2007. 'A Socio-technical Framework for Cyberinfrastructure Design', in *Proceedings of e-Social Science Conference*, Ann Arbor, Michigan, 7–9 October.

Part III

Internal and External Contributions of the Internet

6   Data Observatories

Decentralized Data and Interdisciplinary Research

Thanassis Tiropanis

6.1 Scientific Discourse on the Internet

Publishing and sharing data for the purpose of statistical analysis is an activity that can be traced back to as early as the 16th century in England, with the weekly publication of mortality statistics (Birch 1759). The purpose of those publications was to record deaths during outbreaks of plague in the City of London. It was on those data that John Graunt (1620–1674) was able to provide his 'Natural and Political Observations on the Bills of Mortality' (Birch 1759) – an analysis to understand changes in the population of the City of London and how those might relate to the plague.

The following centuries saw the proliferation of data gathering and statistical analysis in various sectors beyond public health. In recent decades, the Web has made it possible to share data on a large scale and to provide analysis (observations) on data from different resources. However, these activities have been problematic for a number of reasons. First, it is not always desirable for data publishers to make their datasets open to all, as is favored by most Web publishing platforms. There can be legal and ethical reasons behind this choice but it could also be that publishers do not know – and cannot control – the potential use that others can make of those data in a way that is acceptable to them; i.e., it can often be the case that a publisher places significant effort into the curation and publication of a dataset for free, while another party can make profit by using that dataset in a value added online service. Second, it is not easy to discover what data are available, or to establish their relevance to the analysis at hand, and their quality and provenance to that effect. This limits the quality of observations on Web data. Third, there is a barrier in terms of the digital skills required to discover, process and analyze data. While social media have lowered the barrier for publishing content on the Web, there is no endeavor of that scale for sharing data.

As a result, the activity of providing and sharing 'observations' in recent years has been primarily in the hands of individuals with higher

DOI: 10.4324/9781003250470-9

digital skills who have access to datasets gathered by sizeable organizations running online platforms and who are in a position to negotiate and implement the legal and ethical aspects necessary for data gathering and analysis. This not only limits the potential for data-driven innovation but also the potential of research. The decentralization of the data ecosystem to make it more accessible to individuals with varying digital skills promises to benefit society and to further innovation and research.

Data sharing among scientists has been problematic, due to a lack of funding, time, infrastructures and standards when scientists are willing to share their data, while data reuse is hindered by a lack of metadata and related standards.[1] The conundrum of secure sharing has additional aspects. It is not just the technological aspects that can be challenging but also the legal, ethical and social ones, as found in reports cited by Tenopir et al. (2011) for sharing scientific datasets. The complexity of those challenges has led to proposals for Web observatories as socio-technological artefacts for sharing data and observations following a bottom-up approach. The concept of a Web observatory originally started as a means to enable interdisciplinary collaboration in the context of Web science (Tiropanis et al. 2013) and it was subsequently developed as an approach to enable data sharing and analysis of a wider scope, beyond Web science (cf. Tiropanis et al. 2014). In recent years, ethical and legal aspects of data sharing in Web observatories were identified and best practice, as well as a new ethical models, were proposed (Wilson et al. 2016). The technological challenges of real-time data sharing (Tinati et al. 2015) and of sharing databases online (Wang et al. 2017) are two of the technological issues that were further explored. In the meantime, the increasing volume of IoT data, the characteristics of IoT datasets (Siow et al. 2016a) and their potential for analytics present challenges for storage and computation on the cloud and at the edge (Siow et al. 2018), which adds to the technological requirements and challenges of data sharing; personal observatories on lightweight computers at the edge were developed to explore some of those requirements (Siow et al. 2016b).

These developments necessitate one step further in the conceptualization of observatories – namely, the concept of sharing data and observations not only on the Web and on the cloud but also on private cloud and edge deployments. They also necessitate frameworks for the codification and negotiation of access and use of data resources to produce and share observations. In addition, they require mechanisms to establish accountability in data-sharing ecosystems. Regulatory aspects for data and observation sharing platforms need to be discussed, as well as the potential feedback loops of sharing observations within certain groups.

To that end, this paper presents the concept of decentralized data observatories as the next step of Web observatory evolution. Section 6.2

discusses challenges of sharing data and observations with an emphasis on infrastructural and interdisciplinary aspects, while Section 6.3 takes the discussion further, to data observatory concepts and desiderata for data-sharing ecosystems in terms of decentralization, fostering data-driven innovation and supporting evidence-based discourse. Section 6.4 presents a reference architecture for data observatories, discussing how decentralized data-sharing ecosystems based on the proposed reference architecture can support discourse that is inclusive to more stakeholders and perspectives. Finally, lessons from current deployments and guidelines for future data observatories are discussed in Section 6.5.

6.2 The Challenges of Sharing Data and Observations

An imminent data deluge, its significance for different disciplines and the requirements for their storage, curation, annotation and processing were identified in earlier stages of Web development with Hey and Trefethen (2003) making the case for its profound effects on scientific infrastructure. The case for security was also identified in the context of medicine and health, but it was not the focus, since there are significant challenges in the design and operation of digital libraries and middleware to support data-related scientific workflows. The years that followed saw the fulfilment of many aspects outlined in those earlier years. These include scientific data repositories, emerging on organizational, national and international levels, for different disciplines as well as proposals on how to approach the availability and scaling of infrastructures and the documenting of scientific literature on the availability of tools and repositories (Foster, 2005).

Further work is on open science, which is focused on what open science is and how researchers benefit from the availability to datasets. The same benefit goes for the criticism (Mirowski 2018). However, as the availability of such data increased, many of the challenges related to metadata were still unresolved (Tenopir et al. 2011) and there are open challenges for data infrastructures performance and security across interdisciplinary research communities, such as those around the Internet, as a whole. This includes Network science, Web science (Tiropanis et al. 2015), data science (Phethean et al. 2016) and computational data science (Lazer et al. 2009). The work on Web observatories (Tiropanis et al. 2013) aimed to address some of those issues with an emphasis on interdisciplinarity, access control and distributed and decentralized approaches, where data and analytics could be available on different nodes and be managed by different parties who could form groups, subject to sharing common backgrounds (such as ethical frameworks, communities, etc.).

6.2.1 Data Sharing Infrastructures

Having focused on the issue of scaling the storage and compute of datasets, many models of e-science data infrastructures relied on open access, which was not sufficient, while other aspects such as inter-disciplinary support, annotation and metadata to enable discovery and use of datasets by different disciplines and the availability of results/visualizations were still unresolved. Even though privacy approaches were adopted in many proposals, e.g., DataTags (cf. Crosas et al. 2015), there is still no broad agreement on how access to datasets could be insured to remain within groups of authorized researchers to ensure privacy and the observation of legal and ethical requirements.

The challenges of sharing data and analysis for computational social science have been discussed and a self-regulatory regime of procedures, technologies and rules has been proposed; it is argued that the leap from social science to a computational social science is larger than that from biology to a computational biology, given the requirements to request access permission and data encryption (Lazer et al. 2009). In Web science, Web observatories have been proposed to address meaningful engagement with research data (Tiropanis et al. 2013), while, in Internet science, the challenges are of semantic catalogues for data and analytics in order to understand the state and impact of the Internet have been discussed (Wang et al., 2015). From a data science viewpoint, observatories can enable better discovery and engagement with data (Phethean et al., 2016) but there are challenges in negotiating appropriate methodologies. Challenges also include searching and interoperability of data resources (Chen and Yu 2018), for which annotation has shown good potential in many cases (Nowak and Rüger 2010; Willett et al. 2012). There is discussion on issues of privacy, security, consent and trust and how those can be addressed in a wholistic way in the deployment of data-sharing platforms to study socio-technical systems on the Web (Shadbolt et al. 2019) or to deploy data trusts for AI innovation (O'Hara 2019). The reproducibility of AI (Hutson 2018) and bias (Mehrabi et al. 2019; Ntoutsi et al. 2020; Osoba and Welser 2017) give rise to a number of epistemological, ethical and legal concerns.

These concerns can inform top-level design choices for decentralized data observatories in research environments derived from 'The Web Observatory: A Middle Layer for Broad Data', my earlier work in this area.[2] First, it is necessary to provide an architecture where research data would not be shared in a centralized repository but at the research institution that provided and curated them and their analysis; there are legal and ethical reasons for that choice. Second, analysis must cite the data on which it is based, for reasons of accountability and reproducibility, among others. Third, meta-information on the methodology of gathering and analyzing data, as well as on the ethical safeguards, must be available. Fourth, it must be possible to cite data or analysis but also provide different types of access to third parties, and to the research community, for legal and ethical reasons; that requires access control and obfuscation techniques (including summarization and anonymization). Finally, it requires different communities to engage in academic discourse over data and observations, which are necessitated of the means to support that discourse.

The above choices can be further refined as follows:

- *Identifiers*: Identifiers (such as URIs) can be used to identify individuals, communities, data sources and observations in order to determine permitted access and use. Decentralized identifiers can be considered.[3]
- *Meta-information*: Metadata can be used to enable cataloguing, discovery and use of data sources and observations.
- *Access control*: Publishers of data sources and observations control which users can have access to them; they also control how much of the meta-information on those resources can be available to third parties.
- *Consent*: Publishers of data sources and observations can consent to specific uses of those resources and their meta-information by third parties.
- *Portals*: Community- or organization-maintained portals to host meta-information on data and observations are available for sharing by their members, subject to access control and consent. Any community or organization (or even individual) is able to set up their own portal.
- *De-coupling of catalogue from storage*: Portals provide catalogues of resources and meta-information only. Listed resources can be stored in any third-party repository.
- *Inter-portal communication*: Communication between portals maintained by trusted parties is possible. Such communication supports authentication of users, exchange of meta-information on shared resources, search and negotiation of access and use across portals. Building a trusting relationship includes compatibility of legal and ethical frameworks between communities. Standard protocols and schemata are used where possible (e.g., OpenID connect for authentication).
- *Application Programming Interfaces (APIs)*: Where possible, portals provide APIs to support the development and publication of data sources and observations, communication across portals and value added services for their users, to help them monitor how the resources they share are accessed and used.

6.2.2 *Supporting Research Across Disciplines*

The abundance of data has presented opportunities for disciplines beyond physics, biology and other originally envisaged (Hey 2005) to the social sciences and humanities (Chen and Yu 2018), but the need to share best practice, strategies and methodologies, as well as to ensure that ethical issues are observed has been a universal and persistent requirement throughout the years, as discussed in the previous section. In addition, there are wider risks in big data analysis that concern privacy but also its effectiveness in improving decision-making (Chen and Yu 2018), which can be attributed, at least, to the challenges of data-oriented fake news detection and to bias (Ntoutsi et al. 2020).

These emergent contents and procedures point to the significance of involving individual researchers from different disciplines meaningfully and effectively in data infrastructures for research. Further, organizations – and societies with codes of practice, data governance and ethical frameworks – are key stakeholders in data infrastructures and, beyond that, the transparency of sourcing and analysis methods are also significant in scientific discourse across stakeholders (Someh et al. 2019).

Data sharing discussion goes beyond the scientific discourse context, reaching data-driven innovation perspectives (Hemerly 2013) and is subject to critique initially applied to open science (Mirowski 2018).

6.3 The Case for Data and Observation Sharing Ecosystems

The background of data sharing challenges in scientific communities necessitates that data infrastructures support discourse among all stakeholders concerned, by giving them control over what data are shared and how, what processing takes place, what the conclusions of the processing are and with whom they are shared, and what ethical and legal frameworks and safeguards are put in place to support accountability. The concepts of data observatories as data-sharing infrastructures need to take those into account. Giving control to stakeholders, and ensuring ethical and legal frameworks are observed, points to decentralized approaches to those infrastructures, as argued in the previous section. Further, they point to individuals exercising control over how their data are being used in scientific research and beyond, in data-driven innovation ecosystems, and to participating in evidence-based discourse. These points are discussed in more detail in this section and a set of elements that data observatories need to include are proposed.

6.3.1 *Data Observatory Concepts*

From the initial stages of their development, data observatories are conceptualized as communities of people who engage with two types of resources: data sources and observations. The former can be files, data-bases or any query interface that can provide data to be processed by an individual or software. The latter are analytic applications, visualizations, statistical analyses or other forms of data processing obtained from a data source in order to provide an insight or observation. Members of the community can publish and share those resources (Tiropanis et al. 2014) within a community (i.e., within a data observatory). Those who publish resources are called user-publishers or, for simplicity, publishers. Those who access those resources are identified as users. Sharing is subject to the consent of user-publishers to the access and use of those resources by other members of the community. They are also subject to conformance to legal and ethical frameworks set by the community. Sharing across communities, i.e., across data observatories, requires alignment of legal and ethical frameworks and, potentially, additional negotiation to establish the required level of trust. Meta-information or metadata to establish and confirm the provenance, intellectual property, licensing and overall quality of shared resources is also needed within and across data observatory communities. The term metadata is used to describe structured meta-information in a machine-processable format.

On a technological level, each data observatory fulfils its role by providing a catalogue of data sources and observations on a portal, with all the required meta-information. They also involve access control components, access proxies, to ensure that only authorized parties that have negotiated terms of use with the user-publisher get to access published resources. Data sources and analytic applications (observations) can be stored on repositories maintained by the individual user-publisher or on third-party cloud infrastructure. Obfuscation can be applied on accessed data sources or on meta-information on the catalogue; user-publishers can have the freedom to choose how much meta-information on their published resources is available to other users but also to choose the detail, number and frequency of records that can be returned when their data sources are accessed by specific users. Access proxies can use obfuscation to implement access control policies on data sources but also on observations. Access agreements can provide an unambiguous specification of access and use that a publisher grants to another user for a specific resource. Table 6.1 summarizes those elements of data observatories and provides terminology that is used in the rest of this paper.

6.3.2 Decentralization

The potential of data for innovation and research has long been discussed. Data science and Artificial Intelligence (AI) provide tools aiming to fulfil that promise, while different disciplines have explored methods by which to engage with data in meaningful, effective and rigorous ways. The case for decentralization has been adopted in interdisciplinary

Table 6.1 Data Observatory Elements

Element	Description
Data source	A dataset, data stream, database or file shared by a publisher on a data observatory. It involves a user interface (UI) or an application programming interface (API) for obtaining data.
Observation	The result of analysis on data obtained from one or more data sources. It can be a visualization or analytic application, which can be static or dynamically updated. Observations can also involve interaction with the user accessing them and they can be subject to obfuscation by the publisher.
Meta-information	Information or metadata on data sources or observations for the purpose of cataloguing, discovery, attribution, licensing, research use. Vocabularies such as Dublin Core, Schema. org, PROV and DCAT can be used to provide metadata.
Access agreement	An agreement for access to a data source or an observation between a user-publisher and another user. It sets out the access level (access rights, volume and detail of obtained records, frequency of access, use or licensing). It can also specify how much of the meta-information on a published resource can be available to other parties.
Data observatory Portal (DOP)	An often Web-based front end that enables a user to access the catalogue with the meta-information of available data sources and observations; it can also support negotiation with the publisher for access and use of those resources. API access to data sources can be possible for the purpose of developing observations.
User	An agent of a data observatory. Users who publish resources for sharing are identified as *user-publishers* or *publishers*. Users agree to terms for access to the data observatory portal (DOP agreement) and engage in access agreements as users or publishers for access to specific resources.
DOP agreement	An agreement between users of the portal, which can involve access to the catalogue, publication of data sources, publication of observations, availability and use of the portal. This agreement is between each user and the DOP administrator, who acts on behalf of the community or the organization maintaining the portal.
Repository	This is secure storage for data sources or observations, i.e., resources that can be listed in a data observatory. Repositories exist independent of data observatory portals. Access to stored resources is managed by the user-publishers. Those, in turn, can delegate access control to access proxies for users with whom an access agreement has been established in a data observatory.
Access proxy	A component that is responsible for access control to data sources and observations. It can authenticate users and implement access policy on behalf of publishers. It can also enable API access to software used by individuals with the necessary access rights. It monitors and implements the access agreement employing obfuscation if necessary.

(Continued)

Table 6.1 (Continued)

Element	Description
Obfuscators	Software components that can be employed by access proxies in order to perform obfuscation on data obtained by data sources or on observations as per an established access agreement. Such a level of access may concern the number of returned results, number of queries over a certain period, the detail of returned records or frequency of access. Obfuscation can also be employed by the DOP portal so that publishers can establish how much of the meta-information on their listed resources can be available to different parties.

research communities and has informed the conceptualization of data observatories, as discussed in Section 6.2.1 and reflected in the concepts of *access proxies, DOP agreements* and *access agreements* to support access control and consent of parties engaging in data sharing, as discussed in Section 6.3.1. There is also discussion on the limitations of the centralization on the Web in recent years, where a few successful websites have amassed large volumes of data (Berners-Lee 2014).

6.3.3 Data-Driven Innovation

At the same time, technology suites to support re-decentralization have been proposed. The SOLID suite (Mansour et al. 2016) envisages a paradigm where Web service providers (such as online social media providers) do not store user data on centralized repositories. User data are stored and maintained in personal online datastores (PODs) instead. Service providers need to request permission from the users in order to access their POD, instead of storing user data centrally. Web applications rely on access to the PODs of their users, instead of on a central datastore. Users can have control over whether to revoke such access and they have full control of their data at any time. Besides SOLID, the emergence of blockchain can further foster development of decentralized data ecosystems by means of distributed ledgers, the integrity of which can be maintained by a community and supported by consensus-building mechanisms.

Nevertheless, data observatories present an additional challenge: It is not data that need to be shared but observations (e.g., analysis) on those data too. They envisage standard users and two types of publishers: data source publishers and observation publishers. This can be seen as a three-sided market that can foster cross-network effects, where, as more data become available, more observations become available and more people engage with both. However, at the same time, it presents complications when publishers need to control access and use of their resources. There is symmetry, in that both data and observation publishers can control who accesses their resources and for what purpose. However, there is asymmetry, in that observation publishers need to have one agreement with the data publishers for the data resources that they use and a separate agreement with those accessing observations; those two agreements need to be aligned, and accountability needs to be carefully established along the chain of data publisher, observation publisher and observation subscriber.

Even though these concerns seem to be specific to academic research environments now, they could be pertinent to future development of data sharing ecosystems. They necessitate a reference architecture that is specific to data observatories, where both data sources and observations are shared, as opposed to other models where only data sharing is envisaged. Concepts, components and best practice can be developed on such a reference architecture and inform its evolution in the future to support interdisciplinary research but also data-driven innovation.

6.3.4 Evidence-Based Discourse

There is a wider range of stakeholders engaging with data, making appropriate data sharing ecosystems essential beyond research and innovation. The need to promote data and statistical literacy from schools and colleges using appropriate platforms has been argued by Wolff et al. (2016), who also discuss the needs of different types of data-literate citizens (communicators, readers, makers, scientists, etc.) and also make the case for communication between those types to support data-driven inquiry. This makes it even clearer that a data-sharing ecosystem needs to provide content for discourse assuming different levels of literacy and that the most data literate people (e.g., data scientists) need to ensure that what they share is presented appropriately for different data literacy levels. This further supports the separation of the concepts of *data source* and *observation*, as proposed in Section 6.3.1, since it may be necessary to have observations adapted for different types of data literacy, enabling different degrees of engagement of citizens with observations and the data sources that they use.

The socio-technical nature of data infrastructures, which need to accommodate different types of data literacy, has also led to proposals on literacy around data infrastructures themselves, as socio-technical artefacts for creation, extraction and analysis of data (Gray et al. 2018). This will enable identification and discourse around issues such as methodology, bias and remedial actions. It is not a trivial problem to address, but the use of a reference architecture for data infrastructures could provide a foundation to support data infrastructure literacy in this sense.

6.4 A Reference Architecture for Data Observatories

Following up on the concepts presented in Section 6.2, and the elements derived in Section 6.3, abstraction of what is data observatory is presented by means of a reference architecture in this section. This has been informed by concepts reported in Tiropanis et al. (2014) and subsequent development. It describes a number of layers that group components addressing similar areas of concern, where each group relies on the services of the components in the layer below. The reference architecture could help move things forward to the implementation of decentralized data observatories that are more likely to be compatible and easier to harmonize. Figure 6.1 illustrates the four layers that are envisaged:

- *Storage layer*. This is the layer that provides secure storage of data sources and observations by publishers. The storage layer exists independent of data observatories, and it can be instantiated as a database, file system, repository, cloud storage or application server.
- *Access control layer*. This provides secure access to resources in the storage layer, and it can host access proxies that can act on behalf of publishers. An access proxy can employ authentication services to identify users, authorization services to establish access rights to specific resources and obfuscation, as specified in access agreements.
- *Portal layer*. This is the data observatory portal layer, which provides the means to list and discover data sources, observations and relevant metadata by users (subject to the DOP agreement). The portal layer has its own storage of meta-information on listed resources. It can support negotiation of access agreements and it relies on services of the access control layer to ensure that those agreements are respected.
- *Inter-portal communication layer*. Communities of data observatories that have established and maintained relationships of trust can share metadata on resources that they list and enable each other's users to negotiate and obtain access to those resources. Information exchanged between portals is subject to DOP agreements and it can support decentralized searching across data observatories.

Implementations of data observatories such as the Southampton University Web Observatory[4,5] can be seen as instantiations of this reference architecture. A data observatory implementation includes the portal and the access control layers at its core, and it can interface with a number of storage layer providers. Authentication across communities has been implemented using OpenID Connect.[6]

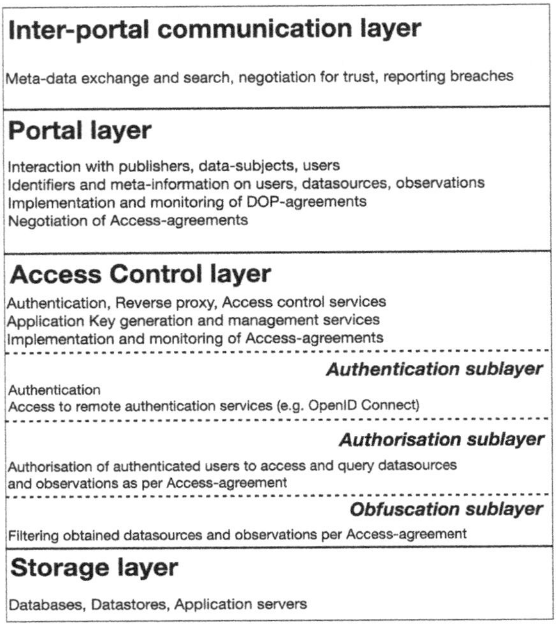

```
┌─────────────────────────────────────────────────────────┐
│ Inter-portal communication layer                        │
│                                                         │
│ Meta-data exchange and search, negotiation for trust, reporting breaches │
├─────────────────────────────────────────────────────────┤
│ Portal layer                                            │
│                                                         │
│ Interaction with publishers, data-subjects, users       │
│ Identifiers and meta-information on users, datasources, observations │
│ Implementation and monitoring of DOP-agreements         │
│ Negotiation of Access-agreements                        │
├─────────────────────────────────────────────────────────┤
│ Access Control layer                                    │
│                                                         │
│ Authentication, Reverse proxy, Access control services  │
│ Application Key generation and management services      │
│ Implementation and monitoring of Access-agreements      │
│ ·······················································  │
│                               Authentication sublayer   │
│ Authentication                                          │
│ Access to remote authentication services (e.g. OpenID Connect) │
│ ·······················································  │
│                                Authorisation sublayer   │
│ Authorisation of authenticated users to access and query datasources │
│ and observations as per Access-agreement                │
│ ·······················································  │
│                                  Obfuscation sublayer   │
│ Filtering obtained datasources and observations per Access-agreement │
├─────────────────────────────────────────────────────────┤
│ Storage layer                                           │
│                                                         │
│ Databases, Datastores, Application servers              │
└─────────────────────────────────────────────────────────┘
```

Figure 6.1 Data Observatory Reference Architecture.

6.5 Deploying Decentralized Observatories

Interdisciplinary research can be seen as a microcosm, where decentralization is essential, building on the structures and practices of academic institutions that predate the industrial revolution and even nation states. The remit of academia as an environment that fosters rigorous discourse has necessitated ways to share and cite resources and methods. The printing press and, more recently, the Internet have provided new ways of sharing data and observations and they inform ethical frameworks and policy. Nevertheless, technical aspects of deploying data observatories are non-trivial. The reference architecture presented in this paper is an abstraction of current architectures that support decentralization in terms of decoupling data (data sources) from analytics (observations) and by fostering mechanisms to negotiate and share those resources with trusted parties. There is no claim that it can guarantee the deployment of a decentralized observatory, but it is argued that it can contribute to the discussion on decentralized data ecosystems and provide content for data observatories that, sharing the same concepts, elements and architectural layers, are more likely to be harmonizable and configurable for interoperability in interdisciplinary research environments.

Beyond research, having discussed the benefits of decentralization for data-driven innovation, it is worth reflecting on readiness to adopt decentralized architectures for data sharing. It can be argued that innovation today relies on data collected by service providers via online platforms or through devices used (and usually paid for) by consumers. AI provides tools to make observations on those data in ways that consumers (and often providers) are not fully aware of. However, those observations have the potential to generate wealth and they also have the potential to change people's lives by informing decision making. Further, it has been argued that the intensity and the mode of user interaction on online platforms not only enables service providers to predict but also to modify behavior (Zuboff 2019). This can lead to vicious circles, where 'observations' by software that is not transparent and cannot be reproduced become self-fulfilling prophecies by means of feedback loops and behavioral modification, which, in turn, could stagnate innovation and lead to dystopian developments. In this light, the case for decoupling data from observations, for making the links between observations and data explicit and for giving people control over access to their data could help avert such developments, but it would not be enough from a scientific viewpoint.

Decentralization requires people to be able to engage with data, methodologies and analysis so that they can make informed decisions about what to share with whom. Individuals need to be able to gain access to the possible knowledge that data can generate (rather than just the raw data) to appreciate its power. Their decision on sharing data needs to be reversible and re-negotiable. Where revenue is generated by their data, they should be able to negotiate a share of that revenue. It is for this reason that the data observatory principle of sharing observations along with data can be empowering. Decentralization and literacy in data sharing ecosystems could pave the way for innovation and wider engagement in academic discourse. Perhaps it is worth noting that, after all, John Graunt, who provided those early 'observations', was not a demographer but a haberdasher by profession.

6.5.1 Technological Considerations

Empowering people to engage in data observatory ecosystems is highly reliant on cloud infrastructure being readily available and affordable to them. Technology suites like that of SOLID can provide the mechanisms for applications to be developed around the data infrastructures that individuals control, as opposed to individuals having their data stored (and often replicated) on infrastructures supplied by application providers. Beyond the opportunity of providing a lower-entry barrier to innovative application developers, this approach will also enable individuals to more meaningfully participate in research that requires access to their data.

Given the decoupling of storage from catalogue – and of data from observation, as presented in the previous sections – data observatories can more readily be deployed around portals, run by parties that bring together communities of individuals engaging in research and innovation on a specific topic, pooling together their data and observations. Credentials that enable the verification of parties running data observatories and the enforcement of data governance principles set by individuals or agreed by the community will be key in providing the necessary trust for engagement and growth in data observatories. The codification and verification of ethical and legal frameworks in data observatories will further support building the required level of trust and setting up inter-portal communication across data observatory communities, benefiting science, innovation and society.

6.5.2 A Policy Perspective

Empowering people to engage in data observatory ecosystems, as argued in the previous paragraphs, needs to address both data protection aspects and digital infrastructure aspects. The former are informed by existing frameworks on data protection, which, however, need to be reconsidered and scrutinized in light of the developments in artificial intelligence and the type of analysis that was not previously possible. The latter, are informed by critical infrastructure policies and shifting policies related to sovereignty and autarky. Policies can vary in different countries and an analysis in a report by EIT Digital on a policy perspective for digital infrastructure and data sovereignty[7] discusses four scenarios depending on the strength of data protection and digital infrastructure control policies.

These policies do not only affect innovation but also research collaboration and scientific discourse. Nevertheless, decentralized observatories have the resilience to adapt to such environments, since the networks that they can build can be reversed or adjusted.

Notes

1 Cf. Tenopir et al. (2011).
2 Cf. Tiropanis et al. (2014).
3 Cf. https://w3c-ccg.github.io/did-spec/ Accessed at 27 July 2021.
4 Cf. https://webobservatory.soton.ac.uk Accessed at 27 July 2021.
5 Cf. https://github.com/webobservatory Accessed at 27 July 2021.
6 Cf. https://openid.net/connect/ Accessed at 27 July 2021.
7 Cf. www.eitdigital.eu/newsroom/news/article/new-report-on-european-digital-infrastructure-and-data-sovereignty/ Accessed at 27 July 2021.

References

Berners-Lee, T., 2014. 'Tim Berners-Lee on the Web at 25: The Past, Present and Future', in *Wired UK*. Available at www.wired.co.uk/magazine/archive/2014/03/webat25/timbernerslee Accessed at 10.5.2016.

Birch, T., 1759. *A Collection of the Yearly Bills of Mortality, from 1657 to 1758 Inclusive*, London: A. Millar.

Chen, S.-H. and Yu, T., 2018. 'Big Data in Computational Social Sciences and Humanities: An Introduction', in S. H. Chen (ed.), *Big Data in Computational Social Science and Humanities, Computational Social Sciences*, Cham: Springer, pp. 1–25. DOI: 10.1007/978-3-319-95465-3_1.

Crosas, M., King, G., Honaker, J., and Sweeney, L., 2015. 'Automating Open Science for Big Data', in *The Annals of the American Academy of Political and Social Science*, vol. 659, pp. 260–273. DOI: 10.1177/0002716215570847

Foster, I., 2005. 'Service-Oriented Science', in *Science*, vol. 308, pp. 814–817. DOI: 10.1126/science.1110411

Gray, J., Gerlitz, C., and Bounegru, L., 2018. 'Data Infrastructure Literacy', in *Big Data & Society*, vol. 5/2. DOI: 10.1177/2053951718786316

Hemerly, J., 2013. 'Public Policy Considerations for Data-Driven Innovation', in *Computer*, vol. 46, pp. 25–31. DOI: 10.1109/MC.2013.186

Hey, T., 2005. 'Cyberinfrastructure for e-Science', in *Science*, vol. 308, pp. 817–821. DOI: 10.1126/science.1110410

Hey, T. and Trefethen, A., 2003. 'The Data Deluge: An e-Science Perspective', in F. Berman, G. Fox, and T. Hey (eds.), *Wiley Series in Communications Networking & Distributed Systems*, Chichester, UK: John Wiley & Sons, pp. 809–824. DOI: 10.1002/0470867167.ch36

Hutson, M., 2018. 'Artificial Intelligence Faces Reproducibility Crisis', in *Science*, vol. 359, pp. 725–726. DOI: 10.1126/science.359.6377.725

Lazer, D., Pentland, A., Adamic, L., Aral, S., Barabasi, A.-L., Brewer, D., Christakis, N., Contractor, N., Fowler, J., Gutmann, M., Jebara, T., King, G., Macy, M., Roy, D., and Van Alstyne, M., 2009. 'SOCIAL SCIENCE: Computational Social Science', in *Science*, vol. 323, pp. 721–723. DOI: 10.1126/science.1167742

Mansour, E., Sambra, A. V., Hawke, S., Zereba, M., Capadisli, S., Ghanem, A., Aboulnaga, A., and Berners-Lee, T., 2016. 'A Demonstration of the Solid Platform for Social Web Applications', in *Proceedings of the 25th International Conference Companion on World Wide Web – WWW '16 Companion*. Presented at the 25th International Conference Companion, ACM Press, Montreal, Quebec, pp. 223–226. DOI: 10.1145/2872518.2890529

Mehrabi, N., Morstatter, F., Saxena, N., Lerman, K., and Galstyan, A., 2019. 'A Survey on Bias and Fairness in Machine Learning', arXiv:1908.09635 [cs].

Mirowski, P., 2018. 'The Future(s) of Open Science', in *Social Studies of Science*, vol. 48, pp. 171–203. DOI: 10.1177/0306312718772086

Nowak, S. and Rüger, S., 2010. 'How Reliable Are Annotations via Crowd-sourcing: A Study about Inter-Annotator Agreement for Multi-Label Image Annotation', in *Proceedings of the International Conference on Multimedia Information Retrieval – MIR '10*. Presented at the International Conference, ACM Press, Philadelphia, PA, p. 557. DOI: 10.1145/1743384.1743478

Ntoutsi, E., Fafalios, P., Gadiraju, U., Iosifidis, V., Nejdl, W., Vidal, M. E., Ruggieri, S., Turini, F., Papadopoulos, S., Krasanakis, E., Kompatsiaris, I., Kurlanda, K. K., Wagner, C., Karimi, F., Fernandez, M., Alani, H., Berendt, B., Kruegel, T., Heinze, C., Broelemann, K., Kasneci, G., Tiropanis, T., and Staab, S., 2020. 'Bias in Data-Driven Artificial Intelligence Systems – An Introductory Survey', in *WIREs Data Mining and Knowledge Discovery*, vol. 10, p. e1356. DOI: 10.1002/widm.1356

O'Hara, K., 2019. 'Data Trusts Ethics, Architecture and Governance for Trustworthy Data Stewardship.' WSI White Paper, n. 1, February, pp. 1–27.

Osoba, O. A. and Welser, W., 2017. *An Intelligence in Our Image*, Santa Monica, CA: RAND Corporation.

Phethean, C., Simperl, E., Tiropanis, T., Tinati, R., and Hall, W., 2016. 'The Role of Data Science in Web Science', in *IEEE Intelligent Systems*, vol. 31, pp. 102–107. DOI: 10.1109/mis.2016.54

Shadbolt, N., O'Hara, K., Roure, D. D., and Hall, W., 2019. *The Theory and Practice of Social Machines*, Published in Lectures Theory and Society. Cham: Springer. DOI: 10.1007/978-3-030-10889-2

Siow, E., Tiropanis, T., and Hall, W., 2016a. 'Interoperable and Efficient: Linked Data for the Internet of Things', in *Paper presented at INSCI 2016, 3rd International conference on Internet Science*, pp. 161–175. DOI: 10.1007/978-3-319-45982-0_15

Siow, E., Tiropanis, T., and Hall, W., 2016b. 'PIOTRe: Personal Internet of Things Repository, '*15th International Semantic Web Conference Posters and Demonstrations Track*, October 2016.

Siow, E., Tiropanis, T., and Hall, W., 2018. 'Analytics for the Internet of Things', in *ACM Computing Surveys*, vol. 51, pp. 1–36. DOI: 10.1145/3204947

Someh, I., Davern, M., Breidbach, C. F., and Shanks, G., 2019. 'Ethical Issues in Big Data Analytics: A Stakeholder Perspective', in *Communications of the Association for Information Systems AIS*, vol. 44/34, pp. 718–747. DOI: 10.17705/1CAIS.04434.

Tenopir, C., Allard, S., Douglass, K., Aydinoglu, A. U., Wu, L., Read, E., Manoff, M., and Frame, M., 2011. 'Data Sharing by Scientists: Practices and Perceptions', in *PLoS One*, vol. 6/6, p. e21101. DOI: 10.1371/journal.pone.0021101

Tinati, R., Wang, X., Tiropanis, T., and Hall, W., 2015. 'Building a Real-Time Web Observatory', in *IEEE Internet Computing*, vol. 19, pp. 36–45. DOI: 10.1109/mic.2015.94

Tiropanis, T., Hall, W., Shadbolt, N., de Roure, D., Contractor, N., and Hendler, J., 2013. 'The Web Science Observatory', in *IEEE Intelligent Systems*, vol. 28, pp. 100–104. DOI: 10.1109/mis.2013.50

Tiropanis, T., Hall, W., Hendler, J., de Larrinaga, C., 2014. 'The Web Observatory: A Middle Layer for Broad Data', in *Big Data*, vol. 2, pp. 129–133. DOI: 10.1089/big.2014.0035

Tiropanis, T., Hall, W., Crowcroft, J., Contractor, N., and Tassiulas, L., 2015. 'Network Science, Web Science, and Internet Science', in *Communications of the ACM*, vol. 58, pp. 76–82. DOI: 10.1145/2699416

Wang, X., Papaioannou, T. G., Tiropanis, T., and Morando, F., 2015. 'EINS Evidence Base: A Semantic Catalogue for Internet Experimentation and Measuremen', in T. Tiropanis, A. Vakali, L. Sartori, and P. Burnap (eds.), *Presented at the Internet Science*, Cham: Springer, pp. 90–99.

Wang, X., Madaan, A., Siow, E., and Tiropanis, T., 2017. 'Sharing Databases on the Web with Porter Proxy', in *Association of Computer Machinery, International World Wide Web Conferences*, Steering Committee, April 2017, pp. 1673–1676. DOI: 10.1145/3041021.3051694

Willett, W., Heer, J., and Agrawala, M., 2012. 'Strategies for Crowdsourcing Social Data Analysis', in *Proceedings of the SIGCHI Conference on Human*

Factors in Computing Systems. Presented at the CHI '12: CHI Conference on Human Factors in Computing Systems, Austin, TX: ACM, pp. 227–236. DOI: 10.1145/2207676.2207709

Wilson, C., Tiropanis, T., Rowland-Campbell, A., and Fry, L., 2016. 'Ethical and Legal Support for Innovation on Web Observatories', in *Proceedings of the Workshop of Data-Driven Innovation in the Web*, May 2016, pp. 1–5. DOI: 10.1145/2911187.2914579

Wolff, A., Gooch, D., Cavero Montaner, J. J., Rashid, U., and Kortuem, G., 2016. 'Creating an Understanding of Data Literacy for a Data-Driven Society', in *Computer Science, Journal of Community of Informatics*, vol. 12/3, pp. 9–26. DOI: 10.15353/joci.v12i3.3275

Zuboff, S., 2019. *The Age of Surveillance Capitalism*, London: Profile Books.

7 Digitization, Internet and the Economics of Creative Industries

Ruth Towse

The 'new' economy signifies the shift to the production and consumption of intangible, ephemeral goods produced in digital form and distributed electronically via the Internet. It has ushered in the concepts of the knowledge and/or creative economy. This begs the question: Do we need new economic analysis to understand these changes?

7.1 Paradigms of the 'New' Economy

What constitutes a paradigm shift in the analysis of change has been hotly debated in economics, as it has in science. In general, it is a break with that which went before – usually a body of thought or a dominant theory. Though the term is often over-used, it can be a useful way of thinking about change – change brought about by the introduction of new technologies, new market structures and players and new business models, as well as other types of change. But are such changes radical enough to cause a paradigm shift in the economic theories used to interpret them, or has the theory simply evolved? I suggest that what has come to be called *platform economics* – the microeconomics of two- and multi-sided markets – calls for this kind of analysis, as it is applied to digitization and use of the Internet in supply and demand for the products of the creative industries.

The creative industries (the arts and cultural industries) have been profoundly affected by the digital revolution. Moreover, they have been subject to the shock of the effect of Covid-19 and many working in these industries are using the Internet to carry them through the economic crisis it has caused. Digitization has altered the fortunes of some cultural organizations, such as museums and opera companies, that were previously viewed as inevitably sclerotic from the economic point of view (and therefore not viable without state support) and it has enabled new market-based industries to emerge, such as video games. The Internet has fundamentally altered participation as well as production techniques and, especially, distribution and sales methods. It has enabled the multitude to display their artistic contributions to the world online, but it has

DOI: 10.4324/9781003250470-10

not essentially altered the craft of professional creativeness and the time needed to produce it. So, the picture is mixed: There has been a shift in the distribution of creative work and in expanded opportunities for reaching the market, but the economics of the initial creation of new works remains more-or-less unchanged.

A pioneering work on the 'Information Age', *Information Rules* (Shapiro and Varian 1999), raised the question of whether a new kind of economics is needed to understand the impact of the Internet. They argued we do not need a new economic approach – 'technology changes, economic laws do not'; economics is general and powerful and can be applied to any form of economic activity. This has been spur to think about the way economics as a discipline is affected by technological change – in this case, the impact of the Internet in the creative industries.

There is a sense in which economics has always done the same thing – analyzed supply and demand, the price mechanism and the market economy – certainly since Adam Smith and some historians of economic thought would go back to Aristotle. The *Wealth of Nations* was written on the cusp of the Industrial Revolution in Great Britain and much of it is as relevant and as readable as when it was written but, equally, most of it is not. It is obvious that the economy has changed over time and that change in technologies and business models alters the context in which economic theory operates. The switch from agrarianism to industrial production made huge changes to people's lives but the basic laws of economics still made sense: There was relative scarcity of resources whose uses were determined by relative costs and prices; some enterprises became large but there were intrinsic limits to expansion. Now that we live in an information-rich digital economy dominated by the Internet, is that still true? It seems that the underlying economic conditions of production have indeed fundamentally changed in some industries. For instance, we need to understand the tendency for the development of ever-larger, ever-growing, dominant multi-national IT corporations. Although microeconomic concepts in industrial organization have apparently shifted, along with changes in technology, we need to ask whether the underlying economic laws are still relevant to the structural change wrought by the digital age and the Internet.

Another aspect is in macroeconomics: National income accounting requires that industries are defined and measured for their contribution to national income and its growth. Consistency is required in national income accounting so that like may be compared with like but it is slow to adapt to structural change in the economy. New categories may be needed for new goods and processes in the online world, in which the boundaries between goods, services and assets are blurred, but change weakens the ability to make 'before and after' comparisons. When the UK introduced the new category of creative industries and the new 'paradigm' of the creative economy, measuring growth before and after that shift in focus became difficult, with the result that unrealistic claims could be made for their economic impact (Towse 2011).

Such changes are basically a matter for government statistical offices but the outcome can be important for policy-making, as data are used for lobbying and rent-seeking purposes, as well as for evaluating government policy. Creative industry lobbies have laid claim to attention from national governments and international organizations because the data they produced on their size and growth purported to show that they are significant sources of national income, both in less developed economies with strong artistic and cultural traditions as well as in post-industrial economies unable to compete with newer manufacturing economies. Initially, at least, the data on which those claims were based were collected and presented in ways that did not conform to best practice national income accounting. That has now changed, however, as government statistical offices now produce the data.

In discussing these developments in relation to government policy and regulation, we need both microeconomic theories about the response of market economies to creativity and innovation and macroeconomic data on the outcome. In the creative industries, copyright law is a favored form of regulation alongside cultural subsidies; can the law adapt to digitization and its accompanying economic change? Does it alter the rationale for subsidies? Is competition law effective in a dynamic multi-national economy of intangible products? These are questions that the analysis raises and attempts to answer.

The chapter continues as follows: Initially, it discusses the use of 'old' economic theories in analyzing the creative industries and 'new' business models. Thereafter, the paper turns to old and new industrial economics applied to the creative industries. Later the focus is on the role of the creator in the creative industries and the impact of digitization in the labor markets for creators and performers. The last section concludes with some final remarks on the impact of the coronavirus in the arts and creative industries.

7.2 Economic change in the creative economy

By the end of the 20th century, reorientation to the terms 'Information Economy' and the 'Knowledge Economy' had taken place, following the decline of, first, heavy industry and then of manufacturing in the 'post-industrial' economy, with its strong reliance on the service economy. The shared characteristics of these terms relate to the production of goods that are intangible, or based on a fundamental input that is intangible, and they reflect the relative size of those sectors in the national income. The OECD (Organisation of Economic Cooperation and Development) in particular espoused the Knowledge Economy and evolved its own research program based on the prominence of human capital, acknowledging the increasing spread of computerization and the economic value of the data it generates.[1] Intangible output may be easily replicated and it effectively becomes a public good, which with the Internet, as the means of distribution, can be disseminated for a very low or zero marginal cost, often requiring protection through intellectual property law (patent, copyright and trademarks) for the appropriation of value.

Of course, neither public goods nor zero marginal cost were new ideas in economics: Public goods have been traced back to Adam Smith, who applied them to defense, law-making and maintenance of the value of currency as collective goods, but applying them to a broad range of the private sector is novel. The problem of zero marginal cost had been highlighted by Dupuit (in the context of the optimum toll for crossing a bridge; see Blaug 1986: 66). The analogy in the digital world is the distinction between the creation of content and its delivery, which is fundamental to understanding the creative economy – the focus of this chapter.

7.2.1 The Creative Economy

The UN, led by UNCTAD (UN Conference on Trade and Development), produced the first *Creative Economy Report* in 2008 that embraced artistic, cultural, scientific and technical creativity. The report identified characteristics that underlie the creative industries. They are:

> the cycles of creation, production and distribution of goods and service that use creativity and intellectual capital as primary inputs; constitute a set of knowledge-based activities, focused but not limited to arts, potentially generating revenues from trade and intellectual property rights; comprise tangible products and intangible intellectual or artistic services with creative content, economic value and market objectives.[2]
>
> (UN 2008, 13)

This definition was based on several prior ways of identifying the creative industries: The term 'cultural industries' (as distinct from 'the arts') had been introduced in France in the 1980s and soon entered the lexicon of UNESCO. Now Cultural and Creative Industries (CCI)' is the tag. Then, in 1998, the UK adopted a creative economy program as a self-conscious means of promoting its perceived leading position in innovation and creativity in the arts and culture.

Before this somewhat unexpected change, the arts in the UK had been 'overseen' by the Office of Arts and Libraries, a very small government department with a very narrow scope, reflecting the UK's policy of 'arm's length' administration of state support via special bodies, such as the Arts

Council, the Crafts Council and, in a different category again, the BBC (which is an independent corporation financed by a license fee). The UK model of dealing with the cultural sector was very different from that in other European countries in which cultural organizations were (and often still are) typically state-owned (by central, regional and/or local government) and staffed by civil servants. The formation of the UK's Department for Digital, Culture, Media and Sport (DCMS) in 1998 involved marshaling together into one 'sector' oversight of these arm's length bodies and elements from other departments of the UK government, such as sound recording and film production, which were previously in the domain of the Department of Trade and Industry. It also included taking over responsibility for copyright from the old Patent Office, which, as discussed below, came to be seen as the lynchpin of the creative industries in the UK (as it did elsewhere). The motive for the transfer was almost entirely political but the realignment established the need for reorganization of industrial classification in the national income accounts, therefore involving economics. These developments, however, pre-dated the widespread adoption of digitization in the creative sector and the new business models that accompanied it that now require realignment or maybe even revolutionary change (Coyle and Mitra-Kahn 2017).

7.2.2 Creative Industries in the UK

The *Creative Industries Mapping Document* published in 1998 by the DCMS used as its definition 'those industries which have their origin in individual creativity, skill and talent and which have a potential for wealth and job creation through the generation and exploitation of intellectual property'. This definition resulted in a list of industries, which has been revised several times. Since 2017, it has comprised advertising and marketing, architecture, craft, product design, graphic design, fashion design, film, TV, video, radio and photography, IT, software, video games and computer services, publishing and translation, museums, galleries and libraries, music, performing arts, visual arts and cultural education, which, together, now identify the UK's Creative Economy.

The data DCMS produced in the first place, however, were not consistent with established national income accounting criteria. There had been no previous category in the accounts that included all those industries and the industries themselves had typically produced their own data on their contribution to GDP (gross domestic product) and its rate of growth, often in order to attract attention. They hyped up their 'economic importance' for rent-seeking purposes in support of the claim for reform of copyright law and to obtain government financial support. In the case of the 'high' arts, there is an added 'moral blackmail' element in the 'merit goods' appeal for greater subsidy. There is indeed an economic case for subsidy or other forms of state support for them, but such data were not – and are not – convincing to those who understand national income accounting. One way or another, the creative industries argued that they should get preferential treatment – the adjective 'creative' is the key here.

Eventually, the Office for National Statistics in the UK stepped in and creative industry data are now routinely handled by it as part of their regular reporting. They show that the creative industries account for around five percent of GDP, rather than the eight percent previously claimed, and that the rate of growth (also at around five percent) has consistently exceeded that of all other sectors of the UK economy.[3] In contrast to the arts, though, it is the IT, software, video games and computer services sub-sector that forms the largest grouping of industries and constitutes nearly two-thirds of the total creative industry GVA (gross value added) in the UK. Copyright law is believed to have played a significant role in their development, raising the question about the economic importance of copyright.

7.2.3 Digitization and Copyright in the Creative Industries

Although several of the creative industries were quick off the mark to adopt digital methods in their production process, they appear to have been taken by surprise by consumers' ability to gain access to their products without payment via the Internet. Copyright piracy led to a huge worldwide campaign following the Napster case in the USA[4] to pressure national governments and international organizations to increase the scope, term, penalties and enforcement of copyright, which has been pursued successively over the last twenty or so years. In the year 2000, the World Intellectual Property Organization (WIPO) instigated a worldwide program of measuring the 'economic value' of copyright, adopting the underlying and unproven assumption that without copyright these industries would either not exist or would contribute much less to GDP.

The involvement of WIPO with the creative industries is another indication of their perceived value in national income creation in both less developed countries and developed ones. It also highlights the hype that goes with the creative economy and the supposedly crucial role of copyright in its size and growth. As new business models have developed, especially online, the level and impact of piracy has been muted, at least in some countries. Overall, business models have proved more successful in displacing piracy than the technological protections measures (TPMs) that were brought in with much fanfare by WIPO at the end of the 1990s. Economics trumped technology![5]

These developments have underscored the role of economic theory in understanding the creative economy, questioning whether a paradigm shift has occurred due to the digital nature of its products and production processes and the role of the Internet in providing access to producers and consumers.

7.3 Old and New Industrial Economics and the Creative Industries

In previous stages of the economy dominated by agriculture, heavy industries and manufacturing, industry size was tied to land, labor and capital as factors of production, in which the individual enterprise (the firm) reached natural limits through diminishing returns to a factor or to the scale of production, with the consequent effect on costs. Costs would initially fall with an increase in inputs and the scale of production but would eventually rise due to inherent bottlenecks as the enterprise got 'too big'. In this 'heavy' economy, producers both created their products and supplied them to the market. Rising costs and inevitable higher prices would eventually kill off demand and the industry would become less profitable, causing investors to switch their capital to other enterprises. Thus, in principle, economic forces would allocate resources efficiently via interaction supply-demand and the price mechanism.

A long-recognized exception to this self-regulating scenario is natural monopoly, in which there are no decreasing returns to scale, even over a very large output, so that marginal costs never rise above average costs. Traditionally applied to utilities such as electricity and telephones – industries with high fixed costs of production and an extensive distribution network – the model has now been applied to cultural goods and services, including theatrical performances and museum exhibitions (Towse 2020a). The problem for the profit-maximizing producer is that marginal cost pricing (the ideal of free marketeers) would not cover long run average costs and so utility industries were typically state-owned or regulated if privatized. The 'classic' natural monopoly is, however, preserved and regulated in order to maximize welfare: The consumer pays the marginal cost (the lowest possible price) and the sunk or fixed cost of the plant is financed by some formula set by regulators or (as in the case of arts organizations) by a subsidy. In that way, regulation works via the price mechanism. Regulation and privatization of utilities have also split up the production of the good from the delivery system – for example, the production of electricity from its distribution network.

There are strong resonances of this model for production in the digital world in which it is important to distinguish the initial production of content (content creation) from distribution, for example, by Internet platforms. Although for many, goods and services fixed or sunk costs have not changed much in the production of the 'prototype', digital products are non-rival and non-excludable when they are made available online, hence the case for copyright to 'privatize' what are, essentially, public goods. On the distribution side, the use of digitization and the Internet has vastly reduced the costs of marketing and delivering content. Digital producers are natural monopolists due to ever-increasing returns to scale in distribution. In the digital creative economy, however, though the model fits, there has been no such established mode of regulation of platforms such as that found in utilities. Only recently has regulation of FAANG (Facebook, Amazon, Apple, Netflix and Google) companies begun to be on the political agenda. Another concern, for example, on the part of the OECD, has been the question of how to regulate multi-sided markets for 'intangible' goods and services (OECD 2018).

While monopoly is universally regarded as a bad thing and competition as a good thing, regulated natural monopoly is an exception, as are copyright and other types of intellectual property monopolies; awarding exclusive control to the creator is justified in terms of their welfare-enhancing role in stimulating creativity and innovation (Landes 2020). Schumpeter famously argued that as (temporary) monopoly enables charging a price greater than marginal cost, it provides the entrepreneur with the incentive to innovate and, therefore, monopoly is necessary for economic growth. He argued that monopoly offers the entrepreneur 'lead time' to capture the returns on their invention before competitors enter the market and compete down the price, coining the phrase 'creative destruction' for the process. Schumpeter's focus was on patents – he omitted to apply it to copyright (Blaug 2005). It should be said, however, that copyright is a weaker form of monopoly than a patent, as it is restricted to copying, per se; 'independent' creation is a defense against the charge of unauthorized copying – something that often crops up, for example, in cases of apparent musical 'theft'.

The theory of creative destruction has been applied to the music industry; it has been argued that record labels were vulnerable to 'destruction' because they failed to adapt to digital downloads and streaming as use of the Internet spread (Handke 2010). Creative destruction requires ever more technological innovation and alerts entrepreneurs who have the funds to invest. Those who already invested in earlier technology (incumbent firms) are likely to be stuck with it for a while, enabling new entrants to stake their claim in the market. Not every innovation is successfully adopted long term, however; in the creative industries, there is some evidence, for instance, that e-publishing is declining while hard copy book sales are rising (Hviid et al. 2019).

Thus, 'traditional' economic models have proved capable of overcoming the problems of financing public goods in these industries – still non-rival but now excludable – and re-establishing economic 'normalcy'. In fact, some industries with public goods characteristics prior to digitization have experienced the process in reverse, enabling the market to work. Over-the-air broadcasting is one example: The signal is accessible to all with equipment (radio and TV sets) that is readily available on the open market. In the past, it was also non-excludable and was financed either by advertisements or by state funding. State broadcasters used to operate a license fee but could enforce payment only by door-to-door monitoring. When TPMs (total productive maintenances) were developed, enforcement costs were reduced but, in many cases, it has been deemed preferable to simply finance public service broadcasting through taxation rather than a license fee (a route not yet adopted in the UK for the BBC, however). Now, though, digital technology has enabled splitting the signal (which is auctioned off to suppliers), enabling private competitors to enter the broadcasting market, with use controlled by digital rights management (DRM) to enforce charging. Thus, what were previously public goods have been privatized via technology. The production of public goods may also be market-based through privately organized collective finance; the growth of crowdfunding, especially for start-ups in the cultural sector, is evidence of the possibility.

A similar story applies to the music industry. Believed post-Napster to be non-viable because of piracy, platforms subsequently reduced free-riding by offering a menu of payment options for streamed products, with one being advertisement-based free supply of music to the consumer and another being a straightforward subscription, both for a huge bundle of titles (Towse 2020a). Netflix offers the same for films and TV shows. In both cases, subscription fees are small and affordable by many consumers (who need to purchase the hardware to access the service, however).

The above economic theories and business models are well-established (some older than others), and they apply to the creative economy. Of course, they were novel in their day and supplanted or challenged 'incumbent' theories. The process of creative destruction also applies to innovation in economic theory! They have, however, come to be dominated by new economic theories that apply, in particular, to Internet-based platforms.

7.3.1 New Economic Theories and the Creative Industries

The paradigmatic shift in analyzing the digital creative industries is the realization that they are subject to both increasing returns to scale and to network effects, mostly, but not only, in respect to the distribution of their products. This combination leads economists to predict that platform-based industries are likely to grow into even bigger natural monopolies.[6] In digital format, there is no limit to the number of consumers or other users who can obtain a perfectly cloned copy of a digital good or join a social media platform, and there are no barriers of entry to creating your own website. There is plenty, not scarcity, in digital output. The limit in the digital economy is the scarcity of time and attention on the part of consumers/users and their numbers – something understood by Herbert Simon with his notion of the 'attention economy', for which *inter alia* he won the Nobel Prize in economics in 1978 (Blaug 1986).

Network effects are demand-side benefits emanating from the uses of online services, such as platforms and social media sites. Unlike 'traditional' external benefits of consumption, though, which are quasi-public goods, network effects can be captured commercially via online markets. The more people who buy the good or use the social media network, the greater the value to others using it and the greater the value derived from it, resulting in a higher willingness to pay. Moreover, the greater the benefits from the network, the more valuable it becomes and so more people have the incentive to join the network, creating exponential growth. Furthermore, multi-homing – the practice of using several online platforms – amplifies these consumption network effects by spreading them across platforms. Thus, technological developments have largely overcome possible negative network effects, such as congestion. Only bandwidth restrictions would limit their scope on the supply side and users' time and attention on the demand side.

In addition, benefits can be captured by platforms during the process of consumers searching for and consuming digital goods. Information provided on people's websites provides online data that can be manipulated electronically to develop individual consumer profiles that have value to others besides the immediate seller (Belleflamme and Peitz 2020). Discrete websites are amalgamated to produce user profiles and information that are sold to other companies for commercial, political and a range of other purposes.

These effects suggest that 'big' is likely to get ever 'bigger': here are no market-based checks to expansion, such as diseconomies of scale. Robots and AI are able to reproduce huge quantities of data at zero marginal cost. The market economy of the Internet therefore cannot self-regulate in the manner perceived in previous technological eras. That is a paradigm shift with strong economic and political implications.

7.3.2 *Platform Economics: Two-/Multi-ided Markets*

The change in economic organization that has evolved quickly in the digital economy is that of two- or multi-sided markets, although, as with so many such developments, there are some precursors. One of the oldest two-sided market models in the creative economy was that of advertisement-based commercial radio (and later, TV), with the broadcaster as the platform: The consumer tolerates the advertisements in order to receive free access to programs. Advertisers buy slots according to the putative audience, thus financing the service. A similar business model exists for ad-financed 'free' digital music distribution via streaming services (Towse 2020a). These models have evolved on the Internet as online platforms offer multiple products and experiences. The key change is complementarity and the fact that network effects produce demand-side externalities, associated especially with online consumption. They are the source of positive benefits to both sellers and buyers of goods and services, as well as to users of online services, such as e-mail and social media sites. As argued above, there are no diminishing returns to scale or long run scale diseconomies to set an eventual limit to their size or number.

Multi-sided markets have several different groups of users on a platform, enabling the use of differential pricing. A classic example is a dating club, which typically attracts more men than women; women are charged a lower subscription fee to encourage them to join. The same market form applies to video game console entrepreneurs in the games industry, who need to appeal to both game developers and players. The platform adopts pricing and other strategies to keep multiple sides engaged, thereby internalizing externalities across the various participants.

Subscriptions to large, bundled repertoires of song, book and film titles have emerged from these market forms, enabling consumers to pay to avoid the advertisements on ad-based sites. These aggregations are no longer in the hands of the 'original' industry, however, but are done by online providers – the Internet platforms. These changes in buying and selling goods and services are novel in today's economy but they were not developed specifically for the digital economy, although the Internet, digitization and mass use of platforms has nurtured them.[7]

Economists have incorporated platform economics and these business models into industrial economics; the methodological question is whether this is a revolutionary paradigm shift or an evolutionary change. I would opt for the latter interpretation, even though there has been a paradigmatic shift in economic organization.

7.4 Supporting Creativity in the Digital Era: The Creator and the Creative Industries

So far, the focus of this chapter has been on industrial organization. The first step in the chain of production in the creative industries (as the DCMS quote above makes clear), however, is the creator – an artist, craftsperson, author or performer. The creative process involves skill and talent, long-term training, experimentation and time, at the end of which a work is produced whose success on the market is uncertain. The greater the novelty, the higher the risk. The risk and cost of production of novel creative work – the time spent and materials utilized – is, in the first instance, borne by creators who then have to engage with cultural entrepreneurs (the 'industry') to get the work to market. Once established, the enterprise may finance further work by the creator, offering them a longer-term contract, with or without an upfront payment, thereby sharing risk to a greater or lesser extent.

Society has several ways of supporting creativity by state action or intervention in the market: publicly funded subsidies, usually to arts organizations, which then make a contract with the creator or performer; philanthropy, which ranges from gifts and sponsorship to prizes, which attract tax remission; and intervention in the market via the law. Copyright law plays a significant role in reducing the risk of others free riding on the cost of creating original works. It protects creators and performers from unauthorized use of their work, thereby offering them an incentive to produce and offer it to the market. Contracts are negotiated between the author (creator/performer) and publisher (record label, film company *et al.*) in a deal that exchanges the right to produce and market a good or service based in the underlying copyright work for a payment. That may be a flat fee or a royalty based on the success of the good embodying the copyright work (the book, the record, etc.) on the market. Generally speaking, the creator or performer, unless she is a superstar, has the weaker bargaining position, which is exacerbated by the extent of concentration in the industry: The fewer enterprises there are, the stronger the concentration and the weaker is the individual creator's bargaining power in relation to the enterprise. The result is that many authors and performers end up with poor deals, often signing away all rights to their work.

The implications of the economic organization of creative enterprises and the relative weakness of the artist in striking the bargain are analyzed in detail for various sectors of the creative industries by Caves (2000). He applies contract theory to the creative industries, building up a picture of contract complexity, from the simple 'handshake' contract between an artist and an art gallery, to those in the film industry involving multiple skills and personnel. Caves' proposition is that the transfer of the rights to creators' and performers' works is required by the industry (rather than a license to use them for a limited time) in order to protect the sunk investment ('sunkenness') from later hold-ups in a sequential chain of production.

All contracts offer terms governing the use of the work and the reward due, though they may take different forms. In some creative organizations, the creator or performer has an employment contract, either permanent or temporary, which means the copyright of work done under that contract belong to the employer. Other contracts range from a transfer of all rights to a work for a single fee (a 'buy-out') to a royalty contract, in which rights may be split up and negotiated separately – for example, the film rights of a book title may be retained by the author in a publishing deal and contracted separately (famously as in the case of author J. K. Rowling, for the Harry Potter books). Different types of contracts offer different incentives and rewards; they also influence, or are influenced by, the structure of the industry. In some creative industries, employment contracts are the norm – for example, for players in an orchestra – while a pop group would have a royalty deal transferring the rights to their performance to a record label. There are many mixed examples as well, but the underlying economic principles are the same.

7.4.1 Impact of Digitization on Creators' Labor Markets

Contractual arrangements and contracts have changed over time in the arts and creative industries; royalty contracts are now the norm in musical and literary publishing, though only since the last century, before which flat fee payments that bought out all rights were common (for example, see Towse 2017 on practice in music publishing). The growth of state subsidy for the arts led to secure employment for performers in established orchestras and opera and dance companies, which, in many European countries, would be part of the state administration and even arms-length arts councils and the like, such as those in the UK, treat the large, national organizations as regular clients with regularly employed personnel.

For those outside the cozy world of this regular support, insecure funding leads to creators and performers having to face many temporary contracts and low incomes, as consistently reported in surveys of artists' labor markets – 'precarity' is the latest term adopted to identify a problem that is as old as the hills. It has been assumed by lawmakers that copyright and performers' rights can support these freelancers, though work in cultural economics has shown that artists, in general, earn well below the national average income. For most, a royalty deal did not (and does not) constitute a substantial part of their income (DiCola 2013; Kretschmer et al. 2019); the exceptions are the relatively few top-earning superstars.

Superstars are likely to dominate artists' labor markets even more in the digital era, due to the expanded markets that the Internet offers. The greatest impact of digitization on artists' earnings, though, has been the transfer of rights to use creators' work to platforms made by the publisher in a deal in which the creator has no say. There has been a great deal of attention on the very low rates of compensation from music streaming by platforms in this context (and similar practices occur in other industries – for example, journalism). The creator or performer has no say in the deal that the publisher or sound recording maker makes with the streaming platform, since most would have transferred the rights to them in the initial contract. Atlhough ad hoc arrangements are made for paying the creator (for example, a streaming rate), these are not contractual arrangements between the creator and the platform (Towse 2020a). Bearing in mind that many initial contracts were (and often still are) for the life of copyright – which varies for authors and performers but often extends, for both, well beyond their lifespan – many find themselves locked-in with only moral rights available to control subsequent use made of their work.[8]

The digital economy has affected the economic organization of creative labor markets as well as product markets in the creative industries; nevertheless, the cost of creation of the underlying content remains basically the same, regardless of the means by which it is delivered to the consumer. It still takes as long to write the book or the song. The presence of social media platforms, however, has opened up the possibility for creators to offer their work directly to the public and many do so – both professional and amateur; in fact, that distinction is becoming increasingly more difficult to maintain. So far, distribution of creative work via digital technologies seems to encourage complementarity rather than substitution between goods in some industries (Bakhshi and Throsby 2014; De la Vega et al. 2020) but there has been little work on the demand side of the equation; adoption of AI might change that, however (Peukert 2019). There has been some research showing that those self-publishers who succeed online are offered contracts with enterprises in the industries, such as publishing and games (Hviid et al. 2019), but the full extent for the practice is not known.

An 'unintended' outcome is that online publication by creators has reduced the search costs (A&R) of enterprises in the creative industries so that they are able to select the ones for promotion that have been proved successful, thus reducing their risk in marketing a work. There is an irony here, in that the early perception of the Internet was that is enabled a 'long tail' of market participation by creators, counteracting the superstar tendency.[9] There is some suggestion that this has led to improved contractual arrangements for those who are approached in this way, as the publisher's risk is reduced. The topic requires further research. The Internet and digitization have therefore altered the status quo in creative labor markets in various ways.

Whatever the impact of these changes on the way artists' labor markets are organized, it does not seem to call for new economic theories to explain the underlying logic of contracting in the creative industries. Moreover, copyright law, which has been adapted to the digital economy, is probably not as significant as an economic incentive as is claimed, since the critical point lies in the contractual arrangements made, rather than the law itself; and, indeed, for most, royalties make up only a small proportion of artists' earnings.

7.5 Final Remarks

This chapter tries to address the extent to which the paradigmatic shift in the Creative Economy due to the adoption of digital technologies and access via the Internet has impacted on the economic theories we use to understand these changes. The question is discussed in the context of the creative industries, which have been considerably affected by these changes, especially during the Covid-19 pandemic. Is economic theory in fact so general and so robust that it can prevail over changes in economic organization, regardless of the technologies adopted, as Shapiro and Varian suggested?

The switch to an economy of plentiful supply in which network effects produce data as a joint product has had a significant impact on many parts of the creative economy. It has enabled sizeable external, even public goods, effects ('spillovers') that need to be internalized by private enterprise. These technological shifts have led to the growth of very large online platforms, which have quickly come to dominate markets in many of the creative industries. It seems that traditional industrial economics has been successfully extended to cope with the digital creative economy and that existing economic concepts have been adapted for analyzing it. Economic theory has responded by evolution rather than by revolution. So, the answer is that Shapiro and Varian were, on balance, correct.

The creative economy consists of both the creation of content and its delivery – and the effects of the Internet and digitization on each have differed. While digital content can be disseminated to vastly expanded markets for virtually zero cost to the supplier, the cost of content creation remains more-or-less unchanged. Many of the activities of the older parts of the creative economy that are building-based, and therefore subject to the limitations of time and space visitors (such as the performing arts and museums) are now available to 'visitors' for online participation via the Internet or through other digital means. I can go to my local arts center in a small town in the UK and pay to watch a narrowcast live performance ('event cinema') from the Metropolitan Opera from New York (for which there is even excess demand), but the performance also can be replicated 'as though live' in various ways. I can 'view' an exhibition from the British Museum in the same way. With improved technology at home (surely only a step away), I could view these events without going out, thereby lifting restrictions of time and space. It remains to be seen what the long-term impact of the lockdown due to Covid-19 has had on the taste for home consumption of creative goods – another possible paradigm shift.

The initial creation of creative goods and services, however, remains fundamentally unchanged, with high fixed costs due to the skill and time required to create them in the first place, regardless of the means whereby they reach the audience. Though digital technologies may be used to assist the creative process, they have not been a substitute for the human effort involved. Artificial Intelligence may change that – some music has been generated that way and 'played' electronically (raising awkward questions for copyright law). Digitization has, so far, not really altered the cost of professional content creation, though it has increased overall supply by others and, perhaps more significantly (though this is rarely discussed), has enabled the resurrection of masses of older content that now competes directly for consumers' attention, the commodity that is ultimately in short supply.

As stated earlier, the limitations of time and attention on the part of consumers may be more significant factors than cost on the production side, even though people are more accustomed to multi-tasking – listening to music while traveling, working or reading and engaging with social media, etc. The market repeatedly demands more content, both new and old. The most serious competition for new work, there-fore, probably comes from the repeated reuse of earlier work. By con-trast to online consumption in the home, though, games enthusiasts (at least, prior to the Covid-19 lockdown) were flocking to large venues and paying entrance fees to observe professional gamers playing video games remotely, with the potential for scarcity of places in the space. Diminishing returns to home consumption may eventually set in – something cultural economists should be considering. The closure of live performance venues and live activities due to the measures taken to control Covid-19 provides a natural experiment to present itself: Having had to participate only online for several months, will consumers return to live events?

All this suggests that we currently live in a dual creative economy, in which the time and skill needed for creativity is more or less unchanged, while the uses to which creative work are put have changed fundamen-tally with digitization and the Internet. At bottom, people still want to experience the arts and be entertained and there are many talented cre-ators and performers ready to offer their services, despite the relatively low average reward. Those services are supplied in both traditional and new modes of delivery.

Some institutional arrangements require root and branch change, how-ever. Given the positive feedback of internalized externalities on the digital economy, regulation of monopoly requires rethinking – for instance, which side of a multi-sided market should be regulated? Governments failed to grasp the changed nature of the information economy, taking a laissez-faire stance, waiting to see how things evolved; by the time it was clear that there are few (if any) limits to the expansion of platforms, it was too late to apply national law and taxation measures to control their unwanted effects, which have to be tackled at the supra-national level. A question also hangs over whether copyright law is capable of adaptation, with some legal scholars believing the change required is so radical as to make it infeasible. Digitization and the Internet have made copyright excessively complicated as it struggles to adapt, so much so that many individual cre-ators for whom its protection is designed fail to either understand or apply it to their work. These types of regulation, however, can only work on an international basis, requiring extensive trade negotiations, which are slow and expensive.

Economics has the means to explain these trends; platform economics is firmly established in the economics agenda. It is, unfortunately, not always adopted in the solution of problems thrown up by new technologies and business models, however (Towse 2020c).

Notes

1 See www.oecd.org/officialdocuments/publicdisplaydocumentpdf/?cote=OCDE/
GD%2896%29102&docLanguage=En.
For recent OECD publications in this area, see www.oecd.org/going-digital/
measuring-the-digital-transformation-9789264311992-en.htm. Accessed at
3.7.2021.
2 These developments are discussed more fully in Towse (2019).
3 There are considerable lags in these data and what is reported here is a 'round
figure'. The arts have been especially badly affected by Covid-19, causing a
huge drop in their level of activity; however, other creative industries, such as
books and games, have increased their revenues.
4 A&M Records, Inc. v. Napster, Inc., 239 F.3d 1004 (9th Cir. 2001).
5 Empirical estimates by Waldfogel (2018) have shown that, overall, the effect
of piracy on output is probably neutral.
6 'Scalability', to use one of the four 's's of the intangible economy according to
Haskel and Westlake (2017): the others are sunkenness, spillovers and synergies. See Towse (2020b) for details in a review of the book.
7 Readers of Jane Austen will remember that in England in the 18th century,
the local Assembly Room acted as a platform, offering dances, concerts, a
library and other facilities for a subscription. The more the 'right sort' of
people took part, the greater the value to other members.
8 Some EU countries, notably Germany and the USA, have 'use it or lose it'
rules, whereby the creator can reclaim the work and possibly renegotiate
terms (see Kretschmer, https://musicbusinessresearch.files.wordpress.com/2012/
04/ijmbr_april_2012_martin_kretschmer_final.pdf).
9 There has been some confusion in the use of the long tail concept: Some use
it to mean that in a statistical distribution of incomes in an occupation, there
are more individuals to be found in the middle range. Others use it in the
sense of the superstar effect, introduced by Rosen (1981), who showed that
increasing market size favored a few top earners.

References

Bakhshi, H. and Throsby, D. 2014. 'Digital Complements or Substitutes? A Quasi-Field Experiment from the Royal National Theatre', in *Journal of Cultural Economics*, vol. 38/1, pp. 1–8.
Belleflamme, P. and Peitz, M. 2020. 'Ratings, Reviews and Recommendations', in R. Towse and T. Navarrete Hernández (eds.), *A Handbook of Cultural Economics*, 3rd edition, Cheltenham, UK and Northampton, MA: Edward Elgar, pp. 466–473.
Blaug, M. 1986. *Great Economists Before Keynes*, Cheltenham, UK and Lyme, USA: Edward Elgar.
Blaug, M. 2005. 'Why Did Schumpeter Neglect Intellectual Property Rights?', in *Review of Economic Research in Copyright Issues*, vol. 2/1, pp. 69–74.
Caves, R. 2000. *Creative Industries: Contracts between Art and Commerce*, Cambridge, MA: Harvard University Press.

Coyle, D. and Mitra-Kahn, B. 2017. *Making the Future Count*, Indigo Prize Entry, September 14. Available at: http://global-perspectives.org.uk/wp-content/uploads/2017/10/making-the-future-count.pdf Accessed at 3.7.2021.

De la Vega, P., Suarez-Fernández, S., Boto-Garcia, D. and Prieto-Rodriguez, J. 2020. 'Playing a Play: Online and Live Performing Arts Consumers' Profiles and the Role of Supply Constraints', in *Journal of Cultural Economics*, vol. 44/3, pp. 425–450.

DiCola, P. 2013. 'Money from Music: Survey Evidence on Musicians' Revenue and Lessons about Copyright Incentives', in *Arizona Law Review*, vol. 55, p. 301. Available at SSRN: https://ssrn.com/abstract=2199058 Accessed at 3.7.2021.

Handke, C. 2010. 'The Creative Destruction of Copyright – Innovation in the Record Industry and Digital Copying'. Available at SSRN: https://ssrn.com/abstract=1630343 or http://dx.doi.org/10.2139/ssrn.1630343 Accessed at 3.7.2021.

Haskel, J. and Westlake, S. 2017. *Capitalism Without Capital: The Rise of the Intangible Economy*, Princeton, NJ: Princeton University Press.

Hviid, M., Izquierdo-Sanchez, S. and Jacques, S. 2019. 'From Publishers to Self-Publishing: Disruptive Effects in the Book Industry', in *International Journal of the Economics of Business*, vol. 26/3, pp. 355–381.

Kretschmer, M., Gavaldon, A. A., Miettinen, J. and Singh, S. 2019. 'UK Authors' Earnings and Contracts 2018: A Survey of 50,000 Writers.' Documentation. CREATe, Glasgow.

Landes, W. 2020. 'Copyright,' in R. Towse and T. Navarrete Hernández (eds.), *A Handbook of Cultural Economics*, 3rd edition, Cheltenham, UK and Northampton, MA: Edward Elgar, pp. 116–128.

OECD 2018. *Rethinking Antitrust Tools for Multi-Sided Platforms*. Available at www.oecd.org/competition/rethinking-antitrust-tools-for-multi-sided-platforms.htm Accessed at 3.7.2021.

Peukert, C. 2019. 'The Next Wave of Digital Technological Change and the Cultural Industries', in *Journal of Cultural Economics*, vol. 43, pp. 189–210.

Rosen, S. 1981. 'Economics of Superstars', in *The American Economic Review*, vol. 71/5, pp. 845–858.

Shapiro, C. and Varian, H. 1999. *Information Rules*, Boston, MA: Harvard Business School Press.

Towse, R. 2011. 'What We Know, What We Don't Know and What Policy-Makers Would Like Us to Know about the Economics of Copyright', in *Review of Economic Research on Copyright Issues*, vol. 8/2, pp. 101–120.

Towse, R. 2019. *A Textbook of Cultural Economics*, 2nd edition, Cambridge: Cambridge University Press.

Towse, R. 2020a. 'Dealing with Digital: the Economic Organisation of Streamed Music,' in *Media, Culture and Society*, vol. 42/7–8, pp. 1461–1478.

Towse, R. 2020b. 'Jonathan Haskel and Stian Westlake: Capitalism Without Capital: The Rise of the Intangible Economy', in *Journal of Cultural Economics*, vol. 44/2, pp. 347–350.

Towse, R. 2020c. 'Figuring It Out: Applying Economics to Copyright Royalty Rates for Streamed Music', in *Review of Economic Research on Copyright Issues*, vol. 17/2, pp. 1–22.

United Nations (UN) 2008. *Creative Economy Report 2008*, Geneva: UNCTAD.

Waldfogel, J. 2018. *Digital Renaissance: What Data and Economics Tell Us about the Future of Popular Culture*, Princeton, NJ: Princeton University Press.

8 Digital Surplus
Labor in the Information Age

Enrique Alonso

8.1 Labor in the Information Age: Upgrading Complexity

The Information Age is the result of a complex interaction between technology and contemporary science. Traditionally, this relationship has been interpreted in a limited scope that somehow left out the social sciences – after all, the Network is a techno-scientific artifact whose native language is that of information theory and computation. The decisions that affect it are adopted by technologists and then implemented in the most diverse fields of human activity. But the Network has demonstrated a capacity for transformation that challenges not only technologists and related scientists but all disciplines. This paper will analyze its impact on contemporary economic and political theory in order to understand what certain doctrines have to say to the Network itself, because not everything that is technically possible is socially and humanly desirable. Specifically, we will focus on the interpretation of new ways of working labor, elaborated from different schools and orientations within the environment of contemporary critical theory.

Critical theory, in recent decades, has developed a considerable effort to interpret digital practices within the Marxist doctrine of labor. The result has been a considerable volume of writings[1] that share a clear purpose: to resignify a certain part of our digital activity as a new form of labor, the so-called *digital labor*. Unfortunately, many of these studies often end up immersed in scholarly debates centered on issues inherited from the Marxist tradition. This shift, understandable from the dynamics of scholar research, ends up obscuring the initial idea by giving little importance to the naïve conception of work that is common in non-academic settings. We will start by taking precisely nothing for granted and by thinking first and foremost about carrying the academic debate into our target: citizenship in the digital age.

For most of us, labor is associated with a generally painful activity linked to subsistence. There will always be those who can argue that labor, *their labor*, does not represent such torture, but even those who are

DOI: 10.4324/9781003250470-11

fortunate enough to enjoy their working life will be able to recognize that it is an activity regulated by schedules and subject to obligations that must be satisfied in order to guarantee survival.

On the other hand, we find the Marxist treatment of *wage labor* within the capitalist system of production. In this case, the psychological component, which associates work with a painful and involuntary activity, is set aside to focus on the existing relationship between the worker and the employer through wages. Therefore, there can be no work without a contract, of whatever kind, and a remuneration, which, in the capitalist system, is made effective through salary. It is true that in post-industrial society[2], wages are not only associated with the reproduction of the labor force but also with consumption capacity, which includes many other goods and services not directly linked to survival. However, I will not dwell on this point, which is more than suspicious after the effects of the financial crisis in 2008.

The Marxist treatment of labor presents an interesting duality that is going to be useful in this study and that, perhaps, because it is obvious to specialists, is not usually treated in this kind of study. It is obvious to the employee that it is not possible to speak of work without a salary, whether it serves to cover the reproduction of his labor force or to provide him with a certain capacity for consumption.[3] The objective of the worker would therefore be to obtain a certain capacity to acquire goods and the means to do so would be wages. Seen from the employer's point of view, things are different. What he seeks in the first place is to obtain the surplus that wage labor generates and the means the employer finds for this is the payment of a salary to his workers. To be more precise, we could say that his purpose is the benefit that would result from subtracting from these gains the expenses linked to the production and financing of his activity. Since we are interested in analyzing the relationship between employee and employer and the way in which each one appreciates a certain activity as work, we will focus on surplus and not on profit.[4]

If a worker carries out an activity for which he receives no remuneration, it will be very difficult for him to understand it as genuine work. It may be a formative activity, for which you will probably have to pay a certain amount, or more commonly leisure, in which case there may or may not be expenses. The employer, on the other hand, will tend to interpret as work, carried out by a third party, any activity that entails a surplus, understanding the salary as a necessary condition for its achievement. As is more than evident nowadays, the entrepreneur will tend to reduce the price he has to pay for this surplus value by acting on all the relevant components, mainly on salary. At the limit, it could be thought that his objective would be to obtain the surplus value, avoiding, if possible, the expenses generated by wages, leaving only the investments in material and the financing as a decrease of his profits. An employer who obtains gains in this way would not fail to consider that there is some work done by third parties, even though he does not have to pay anything for it (cf. Fuchs 2010, p. 191).

From the worker's viewpoint, we can find a similar, but opposite, argument. Again, and always at the limit, an employee could legitimately aspire – if the employer's aspiration is legitimate, too – to earn a salary by minimizing the workforce he employs for it. The optimum would be reached when that force was null and void, by obtaining a salary reward in exchange for a non-existent task. This peculiar kind of worker might think that, to the extent that he obtains a salary, his non-existent activity implies a job for the employer.

Market circumstances prevent such extremes – both the employer's and the employee's – from occurring in practice, but they still represent conceivable extremes. The Marxist doctrine of waged work dissociates the interpretation that the employer and the employee make of labor, since the ends that each one of them pursues are antagonistic. This nuance will be very useful in what follows, since my intention is to clarify the way in which we have come to make present a common form of work, digital labor, in which the worker generates a surplus without receiving a material salary for it, i.e., money.[5]

8.2 Digital Surplus

The analysis of value appropriation in digital media – what I will now call *digital surplus value* – takes place within a varied number of platforms. The best known are the so-called *social networks*, but I believe that this meaning may limit the scope of this study in an obvious way. To solve this problem, I will use the term *social platform*, which includes a greater variety of cases.

Social platforms include, as is obvious, the social networks themselves, i.e., sites such as Facebook, Pinterest, Periscope, etc., but also others, such as search engines, Google, Yahoo, etc., entertainment platforms such as YouTube, Spotify, e-commerce, Amazon eBay, etc., and even dating sites such as Tinder, Meetic and all those that are yet to come. The common characteristic of all of them is that the content that creates the value for which they are visited is generated almost entirely by their own users. Someone might say that rather than talking about users, we should refer to them as genuine *prosumers*,[6] but we understand that the term does not properly represent the situation we are faced with – something that will be discussed in detail later.

Another of the main features of these sites is that they are free of charge. With the exception of platforms geared towards digital commerce, where distribution has a direct cost to the user, the rest are usually presented as apparently free (*gratis-*) platforms – something that is usually perceived with surprise and pleasure by users.

If we now make use of the duality detected in the Marxist concept of waged labor, it is normal that the user of any of these platforms can only conceive of them as part of his leisure and never as environments in which a job takes place, no matter how much he is aware of the benefit that its owners obtain. The question as to the origin of these benefits is somehow canceled by the services that the user obtains free of charge and which are not illogically associated with the healthy exercise of his leisure[7] or the satisfaction of any other need.

The clearest case of this type of false conscience is surely offered by social networks. Any of us would respond, without much hesitation, that the moments spent uploading content to our Facebook profiles, or browsing our friends', are moments entirely devoted to leisure; news stories are increasingly frequent about companies that punish those employees who are caught consulting their profiles on these networks during working hours. The justification that everyone finds for this aggressive measure is usually that it is a loss of working time within working hours. Working time cannot be devoted to leisure; that is all.

But another reason could also be given: the worker who acts in this way not only stops working for his employer but is actually working for a third party. And this is how we got to the very core of this study, because what I intend to show is that it is equally legitimate to understand that the activity that users consider idle actually constitutes working time for the owners of the platform on which they act in each case. To interpret it in this way, it is enough to understand this activity not from our point of view but from the point of view of capital. What we produce with our permanence in such networks is a commodity, measurable not just in information provided, which constitutes the basis of the surplus that the owner of the network obtains.[8] It is about inverting the consideration that users make of their leisure moments in the most diverse digital platforms to understand them and analyze them as work – a peculiar type of work if you want – but work, in short.

The Digital Age is forcing us to produce new frontiers in the definition of human relations and this one is not the least in terms of relevance. I understand that it is perfectly legitimate to resignify digital leisure as a non-salaried form of labor, but labor, in short, insofar as it is the origin of a surplus that the owner of the platform will transfer to the material dimension as soon as he has the slightest opportunity to do so.

One of the main difficulties in making this twist from the traditional perspective is that there is neither a salary nor any obligations that the user must satisfy in a previously agreed manner.[9] As I said at the beginning of the study, work has always been associated with certain obligations, usually painful, in exchange for whose satisfaction the worker receives a salary. None of this is present in this case, and yet there is work.

Let us look first at what we produce. At this moment, digital reality arouses certain suspicions that did not take place years ago, when everything was seen not as a new form of post-industrial capitalism (Ritzer & Jurgenson 2010) but as a positive and desirable consequence of the post-scarcity society. Good examples of the current reluctance are the various data protection laws that have been enacted, or drafted, in recent years. The user – the worker– is increasingly aware that he is passing on valuable information to third parties that may be used without explicit consent. Emphasis is placed on our personal data – the profile data – which are the most keenly protected. But this is not the source of income for digital platforms, no matter how much it is part of it.[10] The real source of this income is a complex combination of production and consumption of information,[11] which guarantees and optimizes both the time spent on these platforms and the number of interactions that the worker carries out on these platforms. Without that work, which results in a digital capital with which networks deal with extreme care, our personal data would be worthless. This would be what the digital worker produces, but what does he get in return? This time, their remuneration cannot be measured in accounting terms – that seems obvious. Rather, we will have to resort to a series of rewards that are difficult to gather in a single category other than that of information.[12]

However, this is not much to say, since the information that the worker receives covers very varied modalities. In the case of Google and the rest of the search engines, it is information offered in the form of links.[13] Social networks, such as Facebook, offer the digital worker in a prominent way. This can take the most diverse forms, text, image, audio and video, as well as links commented or not to third parties (such as Cambridge Analytica). E-commerce focuses on goods of all kinds. These retributions, which the worker appreciates as a free service, are the closest things to his salary, although we can only speak of a non-monetary retribution.

In what follows, I will be concerned, as far as possible, with carefully distinguishing between the digital dimension and the material dimension.[14] The first is basically resolved in information and processing of that information and is difficult to account for. The second rests on the traditional economy and can be measured, though not without difficulty, in economic terms.

What the worker produces translates into information as well as time of use and number of interactions with other information on the platform. In return, he also receives information of the most varied type. The digital capitalist – the legal owner of the platform – also makes an investment in terms of information, which is a genuine *informational investment*. To obtain a better understanding of this point, it may be good to draw on the recent history of the Network. The so-called Web 1.0, or primitive Web, was a service in which the owner of the platform offered all the content. The user[15] only had the opportunity to enjoy them and could not, in any case, contribute with his own content. Web 2.0 puts an end to this balance by allowing the user to provide information, which is currently the essential content of any platform. The informational investment in Web 1.0 could be considered at a maximum since all the content was the responsibility of the owner of the platform. This investment is currently reduced to the management of advertising content, moderation and monitoring, as well as the maintenance of the platform.

In digital terms, the surplus – digital, in this case – would come from deducting the corresponding part of the informational investment made by the platform from the contribution made by the users throughout their activity.[16] Although in the case of Web 1.0 the owner provided all the content, it is now more than evident that the balance is reversed, leaving the media owner as a simple guarantor of a certain solvency and effectiveness of his platform. That balance has shifted the user into the role of a genuine worker and the owner and manager into the role of a capitalist of the Digital Age.

Digital gains accumulate by generating digital capital that will seek its translation into tangible, material capital as soon as possible. The transition, however, is not immediate. It takes a substantial amount of digital capital to generate a formula that allows such translation into physical capital, yet there may be circumstances that make it not easy, or even possible. Many of today's most successful platforms began only by accumulating users, by analyzing and improving how they might be attracted to lend their time to producing and obtaining content (Fuchs 2015, pp. 34–35). When their number, their time of permanence and their production reached acceptable figures[17], monetization was only a matter of time. It was enough to open the doors to advertising and generate the mechanisms that would allow it to be reasonably effective. The evolution of Google or Facebook towards spaces capable of producing an economic return based on advertising began some time ago, through the opening of appropriate locations in their interfaces – locations that have only increased in recent years. Others took longer to do so (Twitter, for example) but have ended up imitating existing formulas that seem to give positive results. Perhaps more difficult is the case of WhatsApp, which, after charging for its service, finally decided to return to the free model, without also including mechanisms of transition or material capitalization.

There does not seem to be a single formula for the transition between digital and material capital; each platform must bid to find its own according to its own characteristics. It is not the same to operate in an environment eminently oriented to desktop use as it to do so in another that is native to the cellular environment, with much less physical space and with capacities limited by the mere size of the devices.

On the other hand, it is also fair to recognize that not every transition from digital to material capital has to be based on the contribution of content generated by its workers (users). Digital surplus can take different forms; some will be based on this contribution, while others will be inclined towards other types of interactions or simply to the exploitation of the time of permanence and, at the limit, we can always find platforms in which it is the owner who contributes the totality of the content. In this case, the surplus value will come from the time workers remain on the platform, being that accumulation of added time, the one that generates the digital capital that will have to be transformed.

8.3 Background and Similarities

The twist between leisure and unpaid digital labor suggested here is not an isolated idea. Our theoretical framework has, so far, been the doctrine of digital labor, but it is not the only one possible. Leisure models that produce surplus through advertising are also not new at all. Leisure is the commodity on which commercial television and advertising space, in general, are based (Ritzer & Jurgenson 2010). The fundamental difference in this case is that the viewer is an anonymous person for the television channel that exploits his time. Only statistical analysis, which is not always reliable, can give an idea of the audience associated with a given space and channel (Fisher 2012, p. 178). On digital platforms, this is not the case. Each one of them has an exact and complete reference of the devices connected at each moment and, insofar as they are services associated with profiles, also of the person who is present at that moment (Bilić 2018). This extends to the user's performance and all interactions that have taken place in that period of time. In the case of television, it can be said that there is undoubtedly a passive – but also anonymous – surplus and therefore it is only measurable in statistical terms. It is the audience that generates that surplus and not an individual profile.

A much closer, and also more difficult, antecedent to analyze has to do with the concept of a prosumer, coined some time ago by Alvin Toffler in his well-known work *The Third Wave*, published in 1980, and which we have already mentioned in this chapter. The idea, later developed by a great variety of authors, sees in the prosumer a consumer to whom certain tasks previously associated with the handling of the merchandise have been delegated (Fuchs 2010, p. 190). Typical examples are the refilling of gasoline by the customer, the selection and bagging in large supermarkets, the assembly of packaged furniture, etc., which we can all recognize without difficulty. In all these cases, however, the user of these services (the consumer) continues to pay a price for the goods he acquires, even if they are half processed, thus retaining their rights over the goods acquired. This reduces its price to the extent that it is the consumer who performs certain tasks that previously corresponded to the producer, but that is all. In the model analyzed here, it is the worker himself who produces practically the entire product without paying for it, and without consequently having the rights that the acquisition of any merchandise generates in the act of purchase.

Ritzer and Jurgenson offers a closer look at the new digital economy. These authors raise, not without a certain timidity, the possibility of facing a new model of expression of capitalism, derived from the reinterpretation of digital leisure as genuine work:[18]

> More specifically, in prosumer capitalism control and exploitation take on a different character than in the other forms of capitalism, there is a trend toward unpaid rather than paid labor and toward offering products at no cost, and the system is marked by a new abundance where scarcity once predominated.
>
> (Ritzer & Jurgenson 2010, p. 12)

Despite the closeness of this statement to our own approaches, Ritzer and Jurgenson refuse to talk openly about labor by reintroducing the psychological components mentioned at the beginning – if there is enjoyment then, there is no work – also ignoring the duality that the Marxist concept of wage labor itself introduces in the point of view of the producer and the employer:

> Thus, we cannot ignore the gains for individuals as reasons for the rise of prosumption. Beyond that, it seems clear that most prosumers seem to enjoy their activities.
>
> (Ritzer & Jurgenson 2010, p. 25)

Another reference that is obligatory in this study focuses on a meme reproduced in various ways and in the most varied contexts: 'if it is free, the product is you'. It seems that this statement could derive from a television interview with Richard Serra and Carlota Fay Schoolman in which they came to argue that 'the product of television, of commercial television, is the audience, and television delivers people to the advertiser' and that 'it is the consumer who is consumed, you are the product of television'. These were certainly provocative statements, but they are centered on an era and a medium that are not the ones analyzed here. However, it has been used many times, especially in the context of data protection. What the phrase openly suggests is that it is the consumer who becomes the commodity of the media. If we reinterpret this statement in the digital medium, what seems to be suggested is that it is our profile data – that is, our digital identity – that would produce the gains that platforms seek to achieve. It is true that the assertion is general enough to fit a multitude of interpretations, but I understand that the term that perhaps produces the confusion that seems most pernicious, to me, is the expression 'merchandise'. Treating us as a commodity invariably focuses our value on a finished product corresponding to our identity. This seems wrong to me. In fact, it might be reminiscent of Marx's polemic against *physiocratic* schools at the origin of political economy. We have already argued that the role of the digital worker is much more varied than the mere constitution of an audience, which also counts in the formation of digital capital, but which is by no means the main value of his activity. Neither is marketing with profile data, however relevant this may be. Without the active production of content by the digital worker, the platforms that are now triumphant in the Internet market would hardly count as anything different from thematic channels from the era of traditional television. It seems evident that our profiles are the objective of the owners of digital platforms – not as finished goods, exchangeable for capital, but as content producers, capable of carrying out effective work in the formation of digital capital.

8.4 Digital Capital and Material Capital

In this section, I am going to clarify some of the key concepts of this work, stating the distinction between the digital and material levels. As I have said before, it is not frequent to find this distinction in the literature on the subject. This will lead us to finally analyze the transition from the digital plane to the material and the mechanisms of surplus production.

What the digital worker produces can be classified into the following categories:

- Time spent on the platform.
- Own contents.
- Interactions with third-party content.
- Profile data both static and those linked to our activity on the Network.

As we have already said, it is not necessary for all of them to be present at the same time for genuine work to take place, although the dominant tendency in recent years has been to increase, as much as possible, the contribution of content on the part of the worker to the detriment of what the owner of the platform had previously had to provide.

What the worker receives as non-monetary remuneration would be summarized as follows[19]:

- Access to the information provided by the platform.
- Access to information provided by other users.
- Interactions with the platform and with other users.
- Personal rewards in terms of recognition and influence.[20]

For this remuneration to be attractive, the owner of the platform must make an informational investment, which basically consists of the following:

- Maintenance of the platform in the Network.
- Access to information, both from the platform and from other users.

- Design and maintenance of attractive interactions for the worker.
- Maintenance of advertising spaces and mechanisms for the promotion of contents and users' profiles.

The difference between what the user contributes and the informational investment made by the platform is what can properly be considered as digital surplus – something that accounts for the amount of information, permanence time and interactions that correspond to the worker himself.

The accumulation of this digital surplus results in the digital capital of the platform. Although something has already been said about the transfer mechanisms between digital capital and material capital, what is sometimes referred to as the *monetization process*, it will be useful to look at it in more detail.

The transfer process can be automatic or take some time. The first possibility is unusual, given the need to accumulate a significant amount of digital capital to activate transfer mechanisms. The history of the Network shows how many of the platforms known today began showing a development model quite far from the material benefit and, in some cases, even contrary to it. But that was a long time ago, when even their own designers conceived digital platforms more as a social and technological experiment than as a genuine social application.

The most common mechanisms for making the transfer to material benefit would be the following:

- Spaces for direct propaganda.
- Promoted links.
- Pay-per-click system.
- Sale of profile data.
- Use of activity data for Big Data.

The gross material capital generated by these transfer mechanisms will have to be reduced by the costs generated by the informational investment charged to the platform, in order to finally account for the material benefit obtained. If we consider the material value of the information produced by the worker and deduct the corresponding part of the material value, also from the information provided by the platform, we would obtain, in this case, the material surplus corresponding to each worker.

This shows the existence of a direct relationship between the material surplus and the digital surplus generated by the workers of the platform, even if it is not necessary a priori. The owner of a platform or social network could decide not to execute any transfer mechanism of those described – or even other possible ones that are yet to come – while retaining a digital capital perfectly available for this purpose.

Twitter maintained this attitude until not long ago, although the pressures and operating costs forced them, already in other hands, to a more than obvious change of attitude.

At this point, one might think that the aim of this work should be to discuss the rights of digital workers over the material surplus generated, but I believe that this debate should start with the digital surplus itself and the work it entails.

At present, some pressure has been exerted on the profile sales mechanism and the use of activity data in Big Data projects. Very little has been said about the other aspects. Recent data protection laws have focused on aspects that do not seriously penalize the owners of large platforms, but perhaps they penalize independent companies dependent on the information that large platforms wish to offer them. The result will be the strengthening of the monopoly of large platforms and the justification for their reluctance to share their data with third parties.

But let us go back to the transfer mechanisms. Any social network that aspires to a digital capitalization capable of material transformation in a given moment must generate mechanisms by which to attract its workers in a growing way and to optimize their work, understood in the variants that we have just specified. There is no single formula and no robust theory about this process but rather the accumulation of a certain number of successful experiences – and failures.

Paradoxically, Facebook is nothing but the successful version of the blogger movement of the early 21st century. In the same way, Google represents a deeper understanding of search engines than Yahoo were able to develop. Thus, it is striking the obvious tendency of Google to promote formulas that curiously remind known strategies adopted by Yahoo. In particular, I am referring to the insistence on relegating the results of searches of your engine to a second level showing instead promoted contents.

The mechanism that each social network finds to optimize its digital capitalization, even before it becomes material, depends on subtle combinations between technical resources, social and individual needs and psychological incentives. This mechanism well deserves the name *platform economy* and refers to those delicate balances that developers, creatives and technicians have to constantly try to maintain for the digital capitalization of the platform. This economy is almost entirely oriented towards the digital realm, i.e., the information that ultimately constitutes the value of the platform. It is also a dynamic formula subject to changes – usually minor and subtle – which seek to optimize the digital work done for the platform. But it also includes an analysis and the consequent response to the actions and preferences of its workers. This kind of apparent freedom of action, to which Ritzer and Jurgenson[21] attach such importance in order to avoid talking about work, is but a natural consequence of the characteristics of the new working environment – the so-called Web 2.0. It should be taken into account that the unforeseen actions sometimes taken by digital workers are, first and foremost, the result of the working framework created by the owners of the platform. It is therefore a dialectical process, characteristic of the informational environment in which the activity takes place. Each new resource tested in the field of platform economics is launched with an intention that must be tested in an environment that is not entirely predictable. The intensive use of the new tool by workers will lead designers to encourage the new resource, redirect it in some way or even eliminate it altogether.[22] However, to see in this process a sign of freedom on the part of the user seems, to me, totally unfocused. Their answers are but a part of the trial-and-error strategy that must necessarily be adopted by developers of the economics of platform in an informational environment. We could talk about freedom if digital workers really had governance over the kind of tools they want to use on the platform, as well as the purposes of the platform, but we know that none of this really takes place. The supposed user independence that so restrains Ritzer and Jurgenson is nothing more than a thorough study of the workers' response to the tools the platform owners offer them in an attempt to optimize their work performance.

Any resource that increases any of the forms of production of surplus described above will be welcomed even at the risk of provoking undesir-able behavior – bullying, fake news, promotion of populism, racism, xenophobia, homophobia, etc. –– which, this time, the platforms will exhibit as a sign of the freedom enjoyed by their users. As we will see, nothing could be further from the true cause of these behaviors.

8.5 Reappropriation: Some Proposals for the Recovery of the Rights of Digital Workers

The Marxist concept of surplus played a decisive role in the collective imagination of workers, progressively transforming them into the work-ing class. The identification of a real job for which nothing is received in return, and whose quantity is also unknown, acted as a spring capable of altering the idea that the worker had of himself as an employee. From considering himself fortunate to receive a payment capable of guarantee-ing his existence and offering him a certain capacity for consumption, he was able to go on to question the legitimacy of the appropriation of part of his workforce through a contract. What I would like to raise by calling here, for various reflections, on the first Industrial Revolution, are our rights on digital surplus, insofar as they involve a form of appropriation of our work in a new environment – that of information. There are two questions that will have to be analyzed; the first relates to whether we really have rights to that unpaid work, which I have called digital work, and the second is about the reappropriation mechanisms that can be made available in case we consider that we are in our right.

It may seem that the answer to the question of whether or not we have rights to the activity we do on social networks is obvious, but the truth is that it is far from being so. First, in the absence of a salary, the user will tend not to identify it as work, but rather as an information-oriented service. As there is generally no direct cost, it will not be identified as a commodity but, at most, as a strange kind of free service. On the other hand, by not consuming, apparently, essential resources that affect our survival, the worker will not feel that he has to claim anything for his activity – he will simply congratulate himself that something so valuable for him does not suppose any cost. I think it is difficult to alter this idea because it is basically correct, always from the exclusive point of view of the digital worker. If we do not reverse the position, putting ourselves in the perspective of the owner of the platform, there will never be anything we can claim as our own. This situation is similar to typical scams in which it is the scammer who thinks that he or she is taking advantage of the situation when, from the scammer's point of view, he or she is just the victim.

For the owner of the platform, as we explained in previous sections, there is a job – a surplus – which finally and, if he wants, will be transformed into material benefits. The legal basis of our claim would therefore be the following: whenever a third party makes a profit from an activity that we have produced, there will be a legitimate claim to the corresponding part of the gains so generated. No matter how little or how much effort we have invested in it, someone has benefited from our activity and therefore we have a right to that benefit. I understand that this is not a simple reasoning, and I am also aware of the way we renounce the rights that do not affect our survival, capacity for consumption or welfare. But we live in an era where it is necessary to revise and even resignify much of our daily lives. I therefore propose that we assume, even as a hypothesis, the existence of a right over the work done on social platforms, insofar as it is a source of benefits of others.

Now that the point has been made, it is time to talk about what these forms of reappropriation can be. I understand that, at present, there are basically two ways of acting on the product of our work: one direct, by claiming a material payment for our activity, and the other indirect, consisting of the socialization of the data generated on the platforms (Fuchs 2010, p. 194). I will begin by discussing the second, which is perhaps more abstract and less striking than the first.

8.5.1 Reappropriation of Data

Much of the digital capital generated by the workers of a platform is not in their profiles, as is usually thought, but in the registration of their activity. This record, stored in databases, allows the systematic exploitation of the worker, generating a new source of commodity: Big Data.

Only now are we in a position to begin to understand what this means in terms of economics and power. There are multiple resources that allow us to exploit this information, obtaining conclusions that extend to practically all areas of daily activity: political and voting trends, tastes, consumption patterns, preferences in leisure, culture, religion and even simple urban and interurban travels. All this information is obtained from the activity records stored in the databases of the social networks.

It is a controversial fact whether or not these platforms use that information – duly processed, analyzed and interpreted – to obtain a direct economic return. Although many of them deny it, given the undoubted impact that the marketing of their privacy could have on their workers, the truth is that the majority at least allows third parties access to their data to perform a task that they themselves do not claim to do (Bilić 2018). The procedure is simple, more than we could ever imagine. All social networks have what, in technical terms, is called an API (application programming interface) – that is, a communication channel that allows third parties to ask questions to its database. Each API has its own syntax and often requires some expertise to operate properly, but nothing that is beyond the reach of any average developer, let alone specialized companies. Access to this interface requires, as a rule, the prior identification of the applicant, at which time the segmentation of the same and the material, economic exploitation of the resources take place. This access is not always the same type: The profiles authenticated by the API can be free, in which case their access permissions are strongly cut, or, on the contrary, they can be paid, giving them different formulas according to the desired level of penetration previously acquired. Obviously, only the platforms themselves retain the full amount of data, as this includes, of course, the personal and private information of their workers' profiles.

The companies or developers who have access to this information could, until recently, process this huge amount of data to produce reports that were made available on the market to the highest bidder or they could, on the contrary, stick to the demand of their customers, generating monitoring and analysis of the contents on demand. Political parties, public corporations and large companies are the usual clients of this type of service.

The recent supranational regulations for data protection have complicated the situation of these companies and developers to the point of making it very difficult to comply with the new requirements. This, however, is not resulting in market exit from the exploitation of data produced by digital workers. On the contrary, the foreseeable – and surely sought – result has been to eliminate from competition small research groups and companies unable to respond to the investments required by the new regulations. It was not about protecting us as users, it was really about concentrating power in fewer hands than the available technology allowed.

Can we demand free access to these data from the great social networks? I am aware that the issue is not minor. Many may think that, in the end, these platforms have the legitimate right of exclusive ownership for data that they themselves take care to maintain, but the truth is that these data are not only theirs, we have produced them and, therefore, we have something to say: we retain a right, even ownership, over them. At this point, it is important to point out that information is not a good like many of those goods we are accustomed to handling. Information is not spent when I consume it or when I share it; it does not belong to the category in which are the goods that we usually consider as such (e.g., water, energy or food) – It is what is called a *non-rival good* (Lessig 2006, pp. 294–295). When I propose free access to data from social networks, I am not depriving them of a good, since my access does not harm theirs or that of others. The only thing I want is to take out of the market a good that I have generated as a digital worker and that I do not want to be exploited for exclusively by those who can pay for it.

I am aware that this proposal can be very abstract, as not many people have the technical expertise to access that information and draw appropriate conclusions. However, my intention is not to propose that we become Big Data analysts. It is enough to reverse the privileged and exclusive position that social networks have acquired over the management of our data. The only way to transform a non-rival good – called, in other contexts, non-fungible – into a rival (or fungible) good is to impose exclusivity in access to that good. It is this maneuver that has been adopted by the major networks provided by the technological infrastructure on which they depend. I very much doubt that getting hold of the value of this digital capital was among the original intentions of most of these information companies. I think, rather, that it was the time and volume of available data that enabled owners, developers and also external researchers to appreciate the material value of the treasured good. It is now, in the present moment, and not at the dawn of the Network, when the game is being played for the ownership of information and, at the moment, we are out of it.

A society in which the information produced by digital workers from the most diverse platforms can be used privately to foresee and determine their movements is, necessarily, a profoundly unequal and potentially insecure society for the vast majority of its members. The measure of releasing the APIs from these platforms by means of public standards – conveniently, subject to industrial standards – would therefore have two effects. The first is to take data analytics off the market, or at least eliminate the privileged position now held by those who can afford the costs of access to the platforms' databases. The second would result not in the elimination of Big Data studies, which I consider impossible, but in their socialization through the participation of various actors with a lower level of resources. It is about letting civil society act by balancing the concentration of immense power in very few hands. Research centers, development collectives, companies – why not – would have access to information and could draw their own conclusions by transferring them to civil proceedings according to their needs and interests. This measure does not exclude a commercial use of the data, it only places it in a real market protected against the monopoly of large data analysis platforms and companies. I think that a world in which resources revert to civil society, even through the laws of the market, is safer than the scenario we are contemplating at the present time with concentrated power in too few hands.

The mechanisms for forcing platforms to take such a drastic step should obviously be civil law and would be limited by the privacy of digital workers' profiles. It is not a question of generalizing some sort of collective espionage but of socializing the data that we actually produce and about which we retain part of the property right. Other collateral measures could include limiting the size of platforms or encouraging the creation of competitive alternatives adapted to these new conditions. I do not think, as utopian as it may be, that these are measures that cannot be rehearsed and, in any case, I do not think they should be put aside for simple shyness or, worse, for sloppiness (Fuchs & Dyer-Witheford 2013, p. 6). The history of the industrial age provides examples of public intervention in the market that are much more aggressive than these – it is enough to look back and, perhaps, not much.[23]

8.5.2 Direct Payment: The Digital Wage

If the reappropriation of the data can seem chimerical to us at the moment, then the possibility of charging for our work on the platforms – of reintroducing the salary in the framework of the digital work – can be simply a nonsensical dream. However, it is not a question now of sticking to what we already have but of analyzing what we are entitled to.

I believe that it has been demonstrated that, from the point of view of the owner of the platform, it is legitimate to talk about work. It is more than evident that there is a material return that has been obtained from the accumulation of digital capital produced by our work. So much for the evidence; it is the digital worker himself who will now have to judge what his perspective is. He can still be considered as a simple user (Alonso 2014) who benefits from a free service and therefore does not consider his rights over the information produced, or he can be recognized as a self-employed digital worker. In this latter case, there are also options. The worker may consider that the non-monetary remuneration he or she receives in terms of information, recognition, etc., is fair and sufficient for his or her work, or he or she may wonder about the destination of his or her surplus value. Only if he does the latter will he put in the right perspective to claim a more substantial return on the digital surplus provided to the owner of the platform.

In this chapter, it is proposed that this return be made in monetary terms – that is, through the direct collection of collaborations. Is this possible? To begin with, keep in mind that this dispute is not new; it has perfectly applicable precedents to the case. Not too many years ago, the mainstream media began offering digital versions of their main titles. The ignorance of the media allowed sites to be created, known as news aggregators – RSS channels – that took advantage of their news to create their own sites, thus diverting the audience to their own media to the detriment, of course, of the original sites. Since then, the fight against RSS channels has been constant under the argument that these pages took, at no cost, information that they had not produced and in the financing of which they did not participate. The argument and the pressures were taken into account and almost all legislation now has measures against this practice, which, in some cases, is heavily penalized.

Are we, the new digital workers, not the primary producers of information for which we do not receive any material payment, because that is what we are talking about and nothing else? The copyright controversy itself has similarities, albeit in a totally opposite direction to that proposed here, as it deliberately ignores the major content producers at the present time who are the digital workers.

But it is a question of analyzing whether it is really possible to obtain a material wage for our activity on social platforms. The management of payments for activity on websites of any kind is not new, nor did it even involve the introduction of particularly disruptive technologies at the time. It was done simply because the resource was already available, but not exploited. Google's pay-per-click mechanism (PPC) is the one that is at the base of much of its business model and is then generalized to the rest of the platforms such as YouTube, Facebook, etc. It is not, therefore, a requirement that implies any technological revolution.

How can we imagine the process of reintroducing wages into digital work? First of all, I would like to make it clear that I do not think that there is an alternative to traditional work here. It is not a question of becoming successful YouTubers or well-known influencers. Will there be those who will succeed? Perhaps, but it is clear that the digital market could not withstand without a serious crisis an enormous number of workers pretending to live off their digital activity. I am thinking, rather, of a very meager – but real – remuneration, linked to a series of mechanisms that are difficult to specify now. The posting of content, the recognition of third party contents, the time of active permanence – all this can be easily monetized, insofar as it is duly registered. It is easy to imagine, although more difficult to substantiate, different ways of rewarding the quality of the contents, or their popularity. I am not opposed to this, and I even see here the opportunity to create mechanisms aimed at favoring quality content and interventions over the mere consumption and production of irrelevant information.

All this activity, duly valued, would pass to the user's account and would be available for management at any time. The formulas here can be varied and not necessarily limited to traditional banking. PayPal-type accounts, as well as legal crypto currencies, could be a viable alternative, at least as much as traditional bank accounts. We are all used to making electronic payments, so I do not see why we could not popularize, if we impose it, the electronic collection – the real realization of the digital wage.

At this point there is a small detail that I have not wanted to comment on until now, which is child labor (Fuchs 2010, p. 192). It is a fact that most digital workers are underage. I admit that it is a problematic aspect to give an opportunity of salaried digital worker with a proposal as disruptive as the present one. But there is no novelty here either; YouTubers and influencers are often minors and there does not seem to have been much controversy surrounding it. What I do find alarming, and very much so, is to recognize that digital work is, today, mostly child labor, something that is presumed and denounced on a material level but which, until now, has gone completely unnoticed on the digital plane.

The very existence of figures like YouTubers deserves some attention in this section. One might think that in some way they already constitute a model of a digital salaried worker, insofar as many of them receive material remuneration through their activity. Admitting that the matter requires further study, I believe that, in general terms, what happens with these workers is that they become shareholders of the company in some way, through participation in the profits of advertising embedded in its contents and other similar formulas. But I think this would require further study.

Unlike in the previous proposal, it does not appear here that legislative pressure can have a direct effect on the problem. Social platforms can invoke a multitude of justifications based on the law in force, to reject any action tending to monetize the work of their digital workers. Ultimately, one can always claim that this, and no other, is the model of its platform and that no legislation can intervene in its business model as long as it is not harmful to general interests. Nor do I see as feasible or realistic a process of awareness-raising of digital workers that takes place in an active struggle such as the one that generated class consciousness in the workers' movement of the nineteenth century. There is no room for nostalgia at this point. I think, rather, of applying the same formula that was at the basis of the success of the platforms for which we now work, its novelty in the face of the alternatives then present. Only the promotion of new platforms based on this new production model, in which workers are paid for their digital work, can prove fortune and challenge the existing monopoly. I do not intend to go so far as to predict the success of such initiatives but the mere possibility of activating them to compete, under new budgets, with existing digital powers already seems, to me, innovative enough not to leave it out of all consideration.

Notes

1 Many of them published in tripleC, a digital magazine edited by Fuchs and Sandoval.
2 This term was coined by J. K. Galbraith and A. Touraine, among others.
3 Marx himself already admits, also as forms of work, the slave labor or the relation of vassalage of medieval farmers.
4 Fuchs (2015), p. 28, moves in a very similar line, although centered on exploitation as a central concept.
5 Fuchs and Sandoval (2014), p. 493, includes here other aspects related to the use of digital media or the prosumer category. We thought it more appropriate to set aside these meanings, which can basically be confused with the standard work implemented by digital means.
6 This term was originally coined by Toffler in *The Third Wave* (Toffler 1980).
7 Fisher (2012), p. 176, presents a lucid view of what social networks bring from the user's point of view.
8 The Google case is developed in depth in Bilić (2018).
9 Fuchs and Sandoval offer in [6] a detailed list of working methods according to whether or not they are employees, as well as other components.
10 This debate is reminiscent of Marx's diatribe against physiocrat schools.
11 Dantas (2017) carries out an interesting study of this complexity from the semiotic point of view.
12 A detailed study of what the digital worker produces and gets in return can be found in Fisher (2012), §3.2.
13 Today, Google offers much more structured and elaborate information than the mere list of links that mark their origins, approaching, paradoxically, the service that used to characterize Yahoo searches.
14 This distinction is not usually dealt with in reference texts within digital labor studies.
15 In this case, the user is assimilated to the case of the spectator, so well analyzed by Smythe (1981).
16 A much more complex view of this concept can be found in Fuchs and Sandoval (2014), pp. 506–507. This complexity comes partly from not separating the material and digital planes in an appropriate way.
17 This is a difficult figure to determine but it is usually a millionaire.
18 By some of their approaches, they approximate what Birkinbine (2018), p. 299, describes as a liberal democrat approach.
19 A different and, in my opinion, more complex classification can be found in Fisher (2012), p. 176).
20 Typical examples would be likes, the rate of followers, mentions, etc.
21 See Ritzer and Jurgenson (2010), p. 21.
22 The case of the *dislike* on Facebook is a good example. So is the evolution of *like* into more complex forms of expression.
23 The privatization of telecommunications market in the 1980s and 1990s is a good example of this, albeit clearly to the contrary.

References

Alonso, E. 2014. *La quimera del usuario*, Madrid: Abada.
Bilić, P. 2018. 'A critique of the political economy of algorithms: A brief history of google's technological rationality', in *TripleC*, vol. 16/1, pp. 315–331.
Birkinbine, B. J. 2018. 'Commons praxis: Towards a critical political economy of the digital commons', in *TripleC*, vol. 16/1, pp. 290–305.

Dantas, M. 2017. 'Information as work and as value', in *TripleC*, vol. 15/2, pp. 816–847.

Fisher, E. 2012. 'How less alienation creates more explotation? Audience labour on social network sites', in *TripleC*, vol. 10/2, pp. 171–183.

Fuchs, Ch. 2010. 'Labour in informational capitalism and on the internet', in *The Information Society*, vol. 26, pp. 179–196.

Fuchs, Ch. 2015. 'The digital labour theory of value and Karl Marx in the age of Facebook, Youtube, Twitter and Weibo', in E. Fisher and Ch. Fuchs (eds.), *Reconsidering Value and Labour in the Digital Age*, London: Palgrave Macmillan, pp. 26–41.

Fuchs, Ch. and Dyer-Witheford, N. 2013. 'Karl Marx @ internet studies', *New Media & Society*, vol. 15/5, pp. 782–796.

Fuchs, Ch. and Sandoval, M. 2014. 'Digital workers of the world unite! A framework for critically theorisingand analysing digital labour', in *TripleC*, vol. 12/2, pp. 486–563.

Lessig, L. 2006. *The Code 2.0*, New York: Basic Books.

Ritzer, G. and Jurgenson, N. 2010. 'Production, consumption, prosumption: The nature of capitalism in the age of the digital prosumer', in *Journal of Consumer Culture*, vol. 10/1, pp. 13–36.

Smythe, D. W. 1981. *Dependency Road: Communication, Capitalism, Consciousness and Canada*, New Jersey: Ablex Publishing Corporation.

Toffler, A. 1980. *The Third Wave*, London: Bantam Books.

Part IV

The Internet and the Sciences

9 Biology and the Internet
Fake News and Covid-19[1]

Wenceslao J. Gonzalez

9.1 Biology in the Context of Relations between the Internet and Science

In the relations between science and the Internet, understood in the broad sense, there are three main groups of sciences. The first is directly connected to one of the three main layers of the network of networks, which are the technological infrastructure (or the Internet in the strict sense), the Web and the cloud computing, along with practical applications and the mobile Internet.[2] The second is the set of sciences that use the network of networks to extend their aims, both in terms of longitudinal novelty and from the perspective of transversal novelty. These include, among others, economics, communication sciences and biology. The third are the sciences that arise from the emerging properties of the network of networks, which generates data hitherto unknown and, furthermore, in massive terms (big data). The sciences of the second group can benefit from the sciences of the third block, mainly from data science, which is a transversal discipline (cf. Cao 2017).

Firstly, this general relationship between biology and the Internet is diversified by the methodological pluralism existing in the biological sciences,[3] which is in line with the ontological levels (micro, meso and macro). Thus, there are methodological differences between biochemical studies, botanical research or ecological investigations. Secondly, 'off-line' research, where there is observation and experimentation of the natural world, is not, in principle, the same as 'on-line' research,[4] which moves in an infosphere rather than a biosphere environment.[5] There are, however, points of convergence, as with the mental experiments (that were so important for Darwinian approach [cf. Lennox 2008a] and nowadays [cf. Lennox 2008b]) and virtual experiments. Thirdly, biological research, especially that related to the network of networks, usually has interdisciplinary elements, due to its connection with other disciplines, mainly from the second and third groups but also from the first.

DOI: 10.4324/9781003250470-13

9.1.1 *Two Main Forms of the Relationship*

There are two main forms of the relationship between biology – *sensu stricto*, biological sciences – and the network of networks. The first is how to relate to the Internet in the broad sense – in particular, the Web – in order to *develop science* through some of its constituent elements (language, structure, knowledge, methods, activity, aims and values).[6] The second is how the network of networks can promote the *dissemination of science*, especially from the multiple expressions of the Web (text, audiovisual, etc.),[7] which can inform scientific content and lead to dismantling fake news on current issues, such as Covid-19 and the use of vaccines.[8]

Between these two forms of relationship, there is bidirectionality, or feedback, in that the publication in a digital version – especially if it is open access – provides knowledge that can help to develop biology in terms of basic science, applied science and application of science.[9] This has been highlighted by the research on the pandemic generated by the outbreak of Covid-19. Basic science information on the genome of the virus was fundamental to be able to work on the molecular biology of the virus.[10] The applied science to generate vaccines aimed at preventing the spread of the virus, using mRNA or other routes, was soon made known and promoted scientific knowledge to solve a concrete, socially relevant health problem, and the biological mutations (alpha, delta, omicron, etc.). Later, the application of science through vaccines by risk groups, age groups, medical history, geographic distribution, number of available doses, etc. was disseminated through various digital media of the network of networks, from web pages to specific apps.

Following the first direction, we have to consider how the network of networks can collaborate with biology as a basic science. This leads to studying how it can offer both explanations and predictions about biological phenomena (such as the virus that generates Covid-19 and the successive mutations known to date). Thereafter, it is possible to analyze how biology uses the network of networks to contribute to predictions (ontological, epistemological and heuristic) about the possible future,[11] which give rise to prescriptions that allow solving concrete problems (as in the case of the pandemic generated by Covid-19). Later, it is worth examining how the network of networks can facilitate a better application of the life sciences to well-defined cases, singling out or personalizing solutions (e.g., to provide biological information that is relevant and reliable for the various groups that are to receive the appropriate vaccine, depending on their circumstances of age, medical history, etc.).

If the second direction, which focuses the attention here, is followed, the question is how the network of networks can lead to two complementary tasks. On the one hand, there is the role of *disseminating reliable knowledge* of biology, so that scientists, organizations and the general public have the knowledge they need (e.g., in hospitals, health centers, advisory committees for public management or policy-making, etc.). On the other hand, there is the task of *eradicating fake news*, which can look directly at the origin of the problem (the sources that generate them),[12] the processes they follow (the transmission routes) or the repercussions to which they give rise (misinformation or disinformation).[13] For this second undertaking, we should attend to social networks, especially the most influential (Facebook, Twitter, etc.), YouTube and also the practical applications (apps) that have been created in the wake of the pandemic.

9.1.2 Instrumental and Substantive Contribution of the Web

The Web can make an instrumental contribution to biology in the two directions outlined above. On the one hand, it can actively collaborate in the *development* of biological sciences either by visualizing or making clearer or more explicit the various complex structures of the biosphere – from the biochemical to the ecological realm – as well as its dynamics. This type of task has a convergence with the tasks that computer sciences do for biological development, which has given rise to a whole branch of knowledge: bioinformatics.[14] On the other hand, the Web is a great instrument in publicizing the various advances in each of the biological sciences, making valuable information accessible to scientists, organizations and the general public.

For years now, in addition to having arisen in a research center (CERN),[15] the Web has been instrumental in the elaboration of scientific knowledge. This can be seen in the fact that the Web allows the transfer of information in real time, thus enabling dynamic interaction between biology researchers from different scientific branches and from diverse laboratories around the world. This is particularly useful for interdisciplinary work on urgent and important problems such as Covid-19, especially the development of vaccines and the preparation of effective treatments for the pandemic. The Web has also been instrumental in making scientific knowledge accessible through the digital edition of publications, especially in open access format and through professional social networks such as ResearchGate or Academia.edu. This has allowed works in biochemistry, molecular biology, genetics, microbiology, botany, zoology or ecology to be consulted from anywhere in the world.

Both cases – scientific production and accessibility of scientific contents – require the technological support of the Internet, in the strict sense, so that, as in other cases, technology is also a conditioning factor.[16] Because the available technology ends up being two things: the condition that makes something possible and the form that the content takes according to the mold used in each case. Thus, technological design modulates the type of scientific design that is used in the network of networks to carry out biological research and for the type of communication (text, audiovisual, etc.) that transmits biological information. In other words, (a) technology is a *conditioning factor* in the development of scientific knowledge in the biological sciences (from biochemistry to ecology) in the network of networks, and (b) technology is the *channel* through which communication must pass for scientists, organizations and the general public (as seen in the information on Covid-19 and vaccines in each country).

Through this instrumental dual task of the Web – where the cloud computing can also play a role and, in principle, but to a lesser extent, so can apps – a substantive contribution is made.[17] This contribution of content affects several aspects. Firstly, the generation of new problems to be investigated, as a consequence of new scientific contributions and their accessibility to the public, as has happened with the Covid-19 problem and vaccines. Secondly, the search for new procedures and methods as new thematic possibilities open up, especially in the field of applied biology and its link with biotechnology. Thirdly, the presence of new results, mainly due to social pressure, as seen for months in the Covid-19 and vaccine research. In the latter case, the problem of the distinction between fast science and rigorous science arose, especially when the former can generate fake news by going too fast.[18]

As a consequence of the substantive contribution, there is a broadening of the thematic possibilities of biology, which may bring either longitudinal novelty or transversal novelty. (I) There are new contributions in the way biological knowledge is presented, so that it becomes more accessible to a larger number of people. (II) The social dimension of biological knowledge is highlighted, both due to its repercussions for humans (as we have seen in the cases of Covid-19 and subsequent vaccines) and for animals, which opens biology to new issues. (III) The presence of the scientific contributions of biology in social networks, on platforms such as YouTube or Youku, makes it possible to measure their social impact, seeing how they permeate the activity of individuals, groups, organizations and institutions.

9.2 Fake News and Covid-19

After placing biology – in the broad sense – in the context of the relationship between the Internet and science, which has led to two forms of relationship between them and the instrumental and substantive contribution of the Web, the next step is to address fake news in the case of Covid-19 and the role of vaccines. This requires first pointing out the problems involved, then highlighting the two forms of fake news: misinformation and disinformation; and, later, indicating the routes to deal with the problems identified.

9.2.1 *Problems Raised*

Regarding Covid-19, there has been – and still are – fake news in each of the relevant aspects mentioned: origin, process and repercussion. (1) On the origin of the virus and the consequent pandemic, there are different versions, some of them diametrically different: (a) They affect the starting point, whether it was purely natural and fortuitous or strictly artificial, within the framework of laboratory work. (b) They concern the initial patient infected, whether it was in Wuhan in the last months of 2019 or whether it existed before that. (c) They deal with the place of emergence of the new virus, whether it was the Chinese city or whether it came from another place.[19]

(2) As for the process of the spread of the virus, which has caused epidemics in each country and has generated a pandemic that has affected millions of people around the world, there are several aspects at stake. The main question is whether this transmission was purely natural – due to social interaction in a globalized world – or whether this diffusion was the result of an initial lack of control, deliberate or not, at the starting point. Then, regarding the spectacular intensity of this spread, it is necessary to determine whether, in initially presenting this pandemic as something similar to a type of influenza, there was irresponsibility on the part of the health authorities at the international or national level, or whether there was genuine ignorance about the virus, whose genome was soon decoded.

(3) About the repercussion of Covid-19 and the subsequent pandemic, there are also a number of aspects to consider: (i) Whether social networks contributed to attenuate or dilute the importance of the contagion of the virus, contributing to demonstrations and public events that increased the actual diffusion of Covid-19; (ii) Whether public management – at the international, national or regional level – was adequate to deal with the problem or whether something published by various digital media was taken as valid and had no basis (the light or non-serious nature of the contagion by this virus, which supposedly would run like a flu; and (iii) The data of infected and deceased in each country – especially in some – vary considerably. Thus, there are, in some cases, very significant differences between official government figures and independent estimates, including those produced by public funding agencies.[20]

9.2.2 *Two Forms of Fake News: Misinformation and Disinformation*

To analyze all of the above, we should distinguish two forms of fake news: misinformation and disinformation. There is misinformation when false information is offered, mainly because of two reasons: (a) the facts have not been established and unfounded statements are conjectured or (b) the hypotheses that are defended should be discarded, due to the complete absence of empirical support. There is disinformation when the information is clearly partial or manifestly biased, either to hide something, to divert attention from what really happened or to mobilize the will of the receivers of the message in a certain direction. This may lead to an attempt to hold another entity or country responsible for what actually happened elsewhere.

Both misinformation and disinformation about Covid-19 – in general, and regarding vaccination, in particular – considered within the complex system of the network of networks (with its three main layers: the Internet, the Web and the cloud computing, along with apps and the mobile Internet),[21] requires attention to three main directions. The first is the scientific side of this complex system. This implies that biology, through the biological sciences directly concerned (such as biochemistry and molecular biology), and other scientific disciplines have to offer epistemic authority against the diffusers of fake news: virus denialists, anti-vaccinationists, etc. The second is the technological facet of the Internet, which has the means to block bots and artifacts designed with Artificial Intelligence that promote misinformation and disinformation. The third is the social dimension, which is articulated in various ways, but which has, in social networks (Facebook,[22] Twitter,[23] etc.) and in audiovisual platforms such as YouTube, an incessant source of misinformation and disinformation.[24]

Prima facie, there may be scientific contribution in several ways. The first is to explain why the virus acts the way it does and to predict in ontological, epistemological and heuristic terms. The second is to solve the problem of contagion itself, where science has to predict to prescribe, leading to a science that can be regulatory. The third is to single out the solutions for the different contexts in which the problem to be solved appears, which, in this case, is in specific patients. Then, after the task of the scientists comes the work of the public managers, who have to ensure that the scientific advice can be carried out in a timely manner.[25] This implies assuming that, in order to reduce or eradicate fake news, it is necessary to have clear limits in terms of barriers or frontiers between what is science and what is not (which includes distinguishing it from proto-science and pseudoscience).[26]

9.2.3 Routes to Deal with the Problems

To face the problems raised and the two expressions of fake news (misinformation and disinformation), the first route looks at the source of the problems, the second deals with the processes and the third is focused on the impact of those problems. Within this threefold framework for the analysis, it seems clear that epistemological, methodological and ontological elements should be considered above all. Thus, the endeavor is a matter of addressing the causes and starting points of fake news (in this case, about Covid-19 and vaccines),[27] the modes and means of dissemination used to spread misinformation or disinformation messages,[28] and the ways to publicize the facts, in order to curb or eliminate the negative impact of misinformation or disinformation on social life.[29]

(I) When the first route looks at the source of the problem, it must address the two main channels that serve as a starting point for fake news. The first is social networks, especially those developed through the Web (although there are now others that follow a different line), which are increasingly participative and personalized.[30] The second comes from the mobile mechanisms that transmit content from social networks and some apps (particularly WhatsApp), capable of creating and spreading fake news about the coronavirus, the need for vaccines and the potential risks of vaccination processes.

Each of the three types of science mentioned above can serve to go to the sources of fake news and, later, try to reduce or eradicate the transmission paths of misinformation and disinformation. Sciences directly related to the network of networks, especially Web science and network science, can deal with the internal mechanisms of communication of the false and the biased. Then sciences such as biology – and, more rigorously, the biological sciences – can address the question of the contents that concern them (in this case, those related to Covid-19 and vaccines). Thereafter, data science can help refute false or biased claims on the basis of knowledge supported by real data.

(II) As the second route to deal with fake news deals with the transmission processes as such, it is necessary to distinguish between elements internal to the processes and external ones, as well as structural, dynamic and pragmatic factors. The internal processes by which fake news about Covid-19 and the role of vaccines reach, in principle, any part of the planet (and the international space station) are those supported by the technological facet, which begins with the Internet (in the strict sense), passes through the Web layer and reaches the third layer (where the cloud computing, apps and mobile Internet are located). The external processes are those that correspond to the social dimension of the Internet, which interact with the internal ones and can be at the service of society,[31] but they can also go in the opposite direction.

Structural, dynamic and pragmatic factors can intervene in order to control the processes of spreading of fake news about biology, in general, and about Covid-19, in particular. First, the technological platform of the Internet has means that structurally allow blocking the transmission of certain content through the network of networks. Second, the social networks of the Web layer (such as Facebook or Twitter) also have dynamic mechanisms to control messages that are launched by their users. Third, there are mechanisms that, in a pragmatic way, can single out or identify practical applications or mobile phones that are used to confuse the population about vaccines or their level of safety.

(III) The third route seeks to reduce the negative impact of fake news. On the one hand, this requires reinforcing the epistemic authority of science and diluting the environmental relativism of those who only see opinions and are unable to arrive at 'facts.' On the other hand, practical solutions are needed that, through labels or other mechanisms, make it possible to reduce the belief of those who are seduced by misinformation and disinformation content or lack prior training to enable them to deal with such false or biased content.[32]

Once again, this route requires attention to the three major elements of the network of networks: the scientific side, the technological facet and the social dimension. In the case of fake news about biology, as has been seen – and continues to be seen – with Covid-19, vaccines and concern for their safety are not just a scientific task. In addition, this is a task that, from the outset, is complex insofar as it initially has a triple philosophico-methodological realm: (a) it has to do with some *contents* (the epistemological aspect), (b) some *ways of delivering the contents* (the methodological component), and (c) they have to be faced so that the *real facts* are known (and, therefore, offer information with true content).[33] This route leads to a task where the scientific side of the network of networks is intertwined with the technological facet and the social dimension. It seems that algorithms, by themselves, are hardly going to solve a problem that concerns what people, in social life, accept (or do not accept) for their daily lives.

9.3 Coda: Scientists as Advisors and Policy-Making

Philosophy of science is increasingly interested in reflecting on the role of scientists as advisors and how they can influence policy-making. In the questions about fake news and Covid-19 related to the Web, when the focus is on the social dimension of the network of networks, there are at least three types of agents involved in the public guidelines, which can be considered stakeholders.[34] First, there are the companies that have a presence on the Web, which can be micro, meso or macro. Second, there are the regulators, which may be international, national or regional institutions, especially governments that think in terms of the short, middle or long term. Third, there are the users of the Web themselves – whether individually, in groups or by creating organizations – that can influence the content on the Web.

Scientists as advisors can offer knowledge about facts, such as those related to Covid-19 and vaccines – and, progressively, about the medicines that are emerging – and they can do so with the three types of actors mentioned above. In the first case, they can do so with companies such as Facebook (within the Meta group) through the independent 'Oversight board'; in the second case, they can do so by collaborating with UN bodies, parliaments or governments; and in the third case, they can do so by offering content individually, through groups of researchers or by creating organizations to inform the public about the characteristics of Covid-19 or existing vaccines and to show that there is no justification for the anti-vaccine movement or that it is absurd to claim that bleach is the key to eradicating the Covid-19 pandemic.

Certainly, there is a transition between scientific knowledge – be it basic, applied or of application – and knowledge management oriented to public action, be it in companies related to the Web (such as Facebook or Twitter), governments or organizations created to influence social networks to eradicate or reduce fake news. Scientists can offer knowledge in terms of probabilities, where the quantitative approach prevails, or as possibilities, where the qualitative allows considering extreme cases, which from the point of view of probabilities have less interest. This advice for the development of public guidelines for action, on the one hand, is based on the epistemic authority of scientists and, on the other hand, is based on the type of values (cognitive, social, etc.) that those who have to formulate public guidelines accept or are willing to accept.

Notes

1 This paper has been written within the framework of research project PID2020-119170RB-I00 supported by the Spanish Ministry of Science and Innovation (AEI).
2 That the design of the Internet, in the broad sense considered here, consists of layers is highlighted in Clark (2018), p. 37. That these layers give rise to a complex system is reflected in Schultze and Whitt (2016). That the complex system as a whole has gone from clearly decentralized to progressively centralized in its layers (especially the third) is indicated in The Economist (2018), p. 5.
3 Methodological pluralism is increasingly accepted in science, in general, and in influential scientific disciplines, in particular. On the characteristics of methodological pluralism, see Gonzalez (2020b).
4 See, for example, Villenueve and Goldberg (2020).
5 On the characteristics of 'infosphere', see Floridi (2014), pp. 40–41, 50 and 59. The topic is discussed in Chapter 2 (pp. 25–58) and pp. 59, 108, 126, 145, 163–166 and 218.
6 These components are found in all sciences. A detailed analysis for the case of economics, involving features with biology, not only in evolutionary economics, is found in Gonzalez (2015b), pp. 11–13.
7 In 2019, the following was published: 'there are currently more than 1.800 billion websites on the Internet' (Mokrý 2019, p. 1577).
8 Cf. Melki et al. (2021).
9 This threefold philosophico-methodological distinction holds true in the case of biology. An analysis as an applied science based on prediction and prescription is to be found in Gonzalez (2015a).
10 The virus SARS-CoV-2 is the contagion that has caused the spread of the disease known as Covid-19, identified in China towards the end of that year.

Previously, other coronaviruses have been known (about seven); two of them have been the cause of other infections already during the 21st century: SARS-CoV and MERS-CoV.

11 Prediction plays an important role in the scientific side of the Internet, which accompanies the technological facet and the social dimension, cf. Gonzalez (2018).

12 Cf. Buchanan (2020).

13 A study of this phenomenon from the perspective of social networks can be found in Shao et al. (2018).

14 Bioinformatics journals publish papers related to the Web, cf. Huse et al. (2014) and Mariano et al. (2016).

15 Cf. Berners-Lee (2000) [1999]. This was the origin of Web science, cf. Berners-Lee et al. (2006).

16 This is evident in the research conducted on the international space station when doing research on human tissue in this non-planetary environment. Cf. ISS U.S. National Laboratory (2021).

17 See, for example, Ren et al. (2010).

18 Cf. King (2021).

19 'Thailand was one of the main destinations for Chinese tourists; the first case outside China was reported there on January 13th, 2020. The first known death from covid-19 outside China occurred in the Philippines', The Economist (2021c), p. 41.

20 'Covid-19 has led to between 7m and 13m excess deaths worldwide, according to a model built by *The Economist*', The Economist (2021a), p. 17.

21 Some practical applications (apps), especially WhatsApp, have been a constant source of fake news about how to prevent Covid-19 or how to cure it after infection. See Moreno-Castro et al. (2020).

22 It is interesting that, in June 2021, Facebook paid for an advertisement in various international media that read the following:
'At Facebook, we're collaborating with European partners to reduce COVID-19 misinformation.
The pandemic has reinforced the importance of collaboration. We're continuing our work with governments, not-for-profit organisations, and researchers globally to support the pandemic response and reduce the spread of COVID-19 misinformation, including:
Partnering with 35 fact-checking organisations covering 26 languages.
Displaying warning screens to highlight possibly incorrect COVID-19 content.
Learn more about our global partnership and find helpful resources at about.fb.com/europe', The Economist, June 12, 2021, full page.

23 Cf. Murayama et al. (2021).

24 Regarding the case of YouTube, see Agence France-Presse (2021).

25 On this issue, see Marco-Franco et al. (2021).

26 Cf. Gonzalez (2016).

27 'One particularly pernicious falsehood is that the mRNA in the vaccine will alter a recipient's DNA. (…) RNA and DNA are different, and mammalian cells have not molecular mechanism for transcribing the former into the latter', The Economist (2020), p. 71.

28 These modes and means have been very numerous in the various phases of the pandemic, mainly since the end of February 2020. Since the vaccines have been available to the public, during the year 2021, 'there has been plenty of disinformation on the safety of inoculations', The Economist (2021b), p. 22.

29 On the main promoters of fake news on Twitter about Covid-19 in the initial months of the pandemic, see Pérez-Dasilva et al. (2020).
30 Cf. Boyd and Ellison (2007), Ellison and Boyd (2013), and Newman (2018).
31 Cf. Gonzalez (2020a).
32 See, for example, what has been done in political science in the case of electoral processes: Clayton et al. (2020).
33 To bring reliable information to the public, the presence of scientists on advisory committees and the epistemic authority of international institutions such as WHO should be increased. Also, supranational political organizations have a role to play in this task, because of the social consequences (such as having millions of people unvaccinated because they are Covid-19 vaccine denialists or abstentionists).
34 Cf. Gonzalez (forthcoming).

References

Agence France-Presse. 2021. 'YouTube Says It Has Removed 1 Million "Dangerous" Videos on COVID', August 25. Available at: https://www.voanews.com/silicon-valley-technology/youtube-says-it-has-removed-1-million-dangerous-videos-covid. Accessed at 27.8.2021.

Berners-Lee, T. 2000 [1999]. *Weaving the Web. The Original Design and Ultimate Destiny of the World Wide Web by Its Inventor*, New York: Harper.

Berners-Lee, T., Hall, W., Hendler, J., Shadbot, N. and Weitzner, D. J. 2006. 'Creating a Science of the Web', in *Science*, vol. 313/5788, pp. 769–771.

Boyd, D. and Ellison, N. B. 2007. 'Social Network Sites: Definition, History, and Scholarship', in *Journal of Computer-Mediated Communication*, vol. 13/1, pp. 210–230.

Buchanan, T. 2020. 'Why Do People Spread False Information Online? The Effect of Message and Viewer Characteristics of Self-reported Likelihood of Sharing Social Media Disinformation', in *PLoS One*, vol. 15/10, pp. 1–33. DOI: 10.371/journal.pone.0239666.

Cao, L. 2017. 'Data Science: Challenges and Directions', in *Communications of ACM*, vol. 60/8, pp. 59–68.

Clark, D. D. 2018. *Designing an Internet*, Cambridge, MA: The MIT Press.

Clayton, K., Blair, S., Busam, J. A., Forstner, S., Glance, J., Green, G., Kawata, A., Kovvuri, A., Martin, J., Morgan, E., Sandhu, M., Sang, R., Scholz-Bright, R., Welch, A. T., Wolff, A. G., Zhou, A. and Nyhan, B. 2020. 'Real Solutions for Fake News? Measuring the Effectiveness of General Warnings and Fact-Check Tags in Reducing Belief in False Stories on Social Media', in *Political Behavior*, vol. 42, pp. 1073–1095.

Ellison, N. B. and Boyd, D. M. 2013. 'Sociality through Social Network Sites', in W. H. Dutton (ed.), *The Oxford Handbook of Internet Studies*, Oxford: Oxford University Press, pp. 151–172.

Floridi, L. 2014. *The Fourth Revolution – How the Infosphere Is Reshaping Human Reality*, Oxford: Oxford University Press.

Gonzalez, W. J. 2015a. 'Prediction and Prescription in Biological Systems: The Role of Technology for Measurement and Transformation', in M. Bertolaso (ed.), *The Future of Scientific Practice: 'Bio-Techno-Logos'*, London: Pickering and Chatto, pp. 133–146 (text) and 209–213 (notes).

Gonzalez, W. J. 2015b. *Philosophico-Methodological Analysis of Prediction and its Role in Economics*, Dordrecht: Springer.

Gonzalez, W. J. 2016. 'Rethinking the Limits of Science: From the Difficulties to the Frontiers to the Concern about the Confines', in W. J. Gonzalez (ed.), *The Limits of Science: An Analysis from 'Barriers' to 'Confines'*, Poznan Studies in the Philosophy of the Sciences and the Humanities, Brill-Rodopi: Leiden, pp. 3–30.

Gonzalez, W. J. 2018. 'Internet en su vertiente científica: Predicción y prescripción ante la complejidad', in *Artefactos: Revista de Estudios sobre Ciencia y Tecnología*, vol. 7/2, 2nd period, pp. 75–97. DOI: 10.14201/art2018717597.

Gonzalez, W. J. 2020a. 'The Internet at the Service of Society: Business Ethics, Rationality, and Responsibility', in *Éndoxa*, vol. 46, pp. 383–412. Available at: http://revistas.uned.es/index.php/endoxa/article/view/28029/pdf

Gonzalez, W. J. 2020b. 'Pragmatism and Pluralism as Methodological Alternatives to Monism, Reductionism and Universalism', in Gonzalez, W. J. (ed.), *Methodological Prospects for Scientific Research: From Pragmatism to Pluralism*, Synthese Library, Cham: Springer, pp. 1–18.

Gonzalez, W. J. forthcoming. 'Stakeholders and the Internet', in M. Bonnafous-Boucher and J. D. Rendtorff (eds.), *New Encyclopedia of Stakeholder Research*, Cheltenham: Edward Elgar.

Huse, S. M., Welch, D. B. M., Voorhis, A., Shipunova, A., Morrison, H. G., Eren, A. M. and Sogin, M. L. 2014. 'VAMPS: A Website for Visualization and Analysis of Microbial Population Structures', in *BMC Bioinformatics*, vol. 15/41, pp. 1–6. DOI: 10.1186/1471-2105-15-41.

ISS U.S. National Laboratory. 2021. 'Engineering Projects to Leverage the ISS National Lab', in *NSF and CASIS Select Three Tissue*, September 8th. Available at https://www.issnationallab.org/iss360/nsf-casis-select-joint-solicitation-tissue-engineering-project/. Accessed at 21.9.2021.

King, A. 2021. 'Fast News or Fake News? The Advantages and the Pitfalls of Rapid Publication through Pre-print Servers during a Pandemic', in *EMBO Reports*, vol. 21/6, pp. 1–4. DOI: 10.15252/embr.202050817.

Lennox, J. G. 2008a. 'The Evolution of Darwinian Thought Experiments', in W. J. Gonzalez (ed.), *Evolutionism: Present Approaches*, A Coruña: Netbiblo, pp. 63–76.

Lennox, J. G. 2008b. 'Thought Experiments in Evolutionary Biology Today', in W. J. Gonzalez (ed.), *Evolutionism: Present Approaches*, A Coruña: Netbiblo, pp. 109–120.

Marco-Franco, J. E., Pita-Barros, P., Vivas-Orts, D., Gonzalez de Julian, S. and Vivas-Consuelo, D. 2021. 'COVID-19, Fake News, and Vaccines: Should Regulation Be Implemented?', in *International Journal of Environmental Research and Public Health*, vol. 18/2. DOI: 10.3390/ijerph18020744.

Mariano, D. C. B., Pereira, F. L., Aguiar, E. L., Oliveira, L. C., Benevides, L., Guimarães, L. C., Folador, E. L., Sousa, Th. J., Ghosh, P., Barh, D., Figueiredo, H. C. P., Silva, A., Ramos, R. T. J. and Azevedo, V. A. C. 2016. 'SIMBA: A Web Tool form Managing Bacterial Genome Assembly Generated by Ion PGM Sequency Technology', in *BMC Bioinformatics*, vol. 17 (Suppl 18). DOI: 10.1186/s12859-016-1344-7.

Melki, J., Tamin, H., Hadid, D., Makki, M., El Amine, J., Hitti, E. and Gesser-Edelsburg, A. 2021. 'Mitigating Infodemics: The Relationship between News Exposure and Trust and Belief in Covid-19 Fake News and Social Media Spreading', in *PLoS One*, vol. 16/6, pp. 1–13. DOI: 10.1371/journal.pone.0252830.

Mokrý, S. 2019. 'Assessing Web Surface Credibility by Generation Y: A Q Methodological Study', in *Acta Universitatis Agriculturae et Silviculturae Mendelianae Brunensis*, vol. 67/6, pp. 1577–1585.

Moreno-Castro, C., Vengut-Climent, E., Cano-Orón, L. and Mendoza-Poudereux, I. 2020. 'Estudio exploratorio de los bulos difundidos por WhatsApp en España para prevenir o curar la COVID-19', in *Gaceta Sanitaria*, July 25th, p. 9. DOI: 10.1016/j.gaceta.2020.07.008.

Murayama, T., Wakamiya, S., Aramaki, E. and Kobayashi, R. 2021. 'Modeling the Spread of Fake News on Twitter', in *PLoS One*, vol. 16/4. DOI: 10.371/journal.pone.0250419.

Newman, M. 2018. *Networks*, 2nd edition, Oxford: Oxford University Press.

Pérez-Dasilva, J. A., Meso-Ayerdi, K. and Mendiguren-Galdospín, T. 2020. 'Fake news y coronavirus: Detección de los principales actores y tendencias a través del análisis de las conversaciones en Twitter', in *Profesional de la Información*, vol. 29/3, pp. 1–22 (online version). DOI: 10.3145/epi.2020.may08.

Ren, J., Williams, N., Clementi, L., Krishnan, S., and Li, W. W. 2010. 'Opal Web Services for Biomedical Applications', in *Nuclei Acid Research*, vol. 38, Web server issue, pp. W724–W731. DOI: 10.1093/nar/gkq503.

Schultze, S. J. and Whitt, R. S. 2016. 'Internet as a Complex Layered System', in J. M. Bauer and M. Latzers (eds.), *Handbook on the Economics of the Internet*, Cheltenham: Edward Elgar, pp. 55–71.

Shao, Ch., Hui, P.-M., Wang, L., Jiang, X., Flammini, A., Menczer, F., Ciampaglia, G. L. and Barrat, A. 2018. 'Anatomy of an Online Misinformation Network', in *PLoS One*, vol. 13/4, pp. 1–23. DOI: 10.371/journal.pone.0196087.

The Economist 2018, 'More Knock-on Than Network. How the Internet Lost Its Decentralised Innocence', *Special Report: Fixing the Internet*, vol. 427/9098, June 30, pp. 5–6.

The Economist 2020. 'Vaccine Safety. An Injection of Urgency', Section *Science and Technology*, December 5, p. 71.

The Economist 2021a. 'Counting the Death', Section *Briefing: The Covid-19 Pandemic*, May 15, pp. 17–19.

The Economist 2021b. 'Covid-19 in Europe. Hot Shots', Section *Europe*, June 5, pp. 21–22.

The Economist 2021c. 'The Covid-19 Pandemic. The Shoe Drops', Section *Asia*, September 11, pp. 41–42.

Villenueve, P. J. and Goldberg, M. S. 2020. 'Methodological Considerations for Epidemiological Studies of Air Pollution and the SARS and COVID-19 Coronavirus Outbreaks', in *Environmental Health Perspectives*, vol. 128/9, pp. 1–13. DOI: 10.1289/EHP7411.

10 The Novelty of Communication Design on the Internet

Analysis of the Snapchat Case from the Sciences of the Artificial[1]

Maria Jose Arrojo

10.1 The Sciences of Design as an Epistemological and Methodological Framework for the Analysis of Communicative Designs on the Internet

The sciences of the artificial offer novel philosophical approaches to communicative designs related to the Internet – those being developed today and those that may be developed in the near future. These sciences provide a solid scientific framework for the network of networks. Based on these disciplines, and with the support of technology, communicative designs on the Internet can achieve new aims – some of them, until now, unimaginable – and do so in an increasingly accessible way.

To analyze the new communicative designs on the Internet, it is useful to reflect on three different aspects of the Internet: the scientific side, the technological facet and the social dimension (Gonzalez 2018a, pp. 77–79; and Gonzalez 2020a). These three aspects of the Internet are interrelated and complementary. Firstly, there is the scientific side, which includes all the sciences that serve as the basis for the development of the Internet as a technological platform.[2] Secondly, the technological facet provides the support that configures the network of networks and has allowed the growth of the Internet infrastructure over the years. Thirdly, there is the social dimension, which complements the scientific and technological aspects of the Internet.

Within this setting, there are three types of sciences related to the Internet, pointed out by W. J. Gonzalez (2018a, p. 95). The first group is at the service of the Internet, as is the case of Web science, Network science and specific Internet science. The second is made up of scientific disciplines that use the network for their developments in new directions, such as communication sciences, information science or economics. The third comes with the disciplines that deal with the emerging properties of the network, such as Data science.

This paper analyzes how the Internet is enabling the development of communication sciences. The continuous advancements in the field

DOI: 10.4324/9781003250470-14

of communication depend mainly on the artificial component of communication. This factor expands the capabilities of human beings to reach new goals. These advancements in the field of communication are directly related to the progress of the Internet itself.

In this regard, communication is a scientific discipline based on designs, which are now frequent, and expand human communicative possibilities. Thus, from an epistemological point of view, communicative phenomena are based on designs that have well-defined aims and specific processes, which seek certain results. From the methodological point of view, the internal development in terms of aims, processes and results requires the external support of information and communication technologies (ICT) (Arrojo 2015). The analysis of these designs will help us to understand the complexity and novelty of the communicative phenomena that develop in this new artificial scenario.

10.1.1 Types of Communicative Designs on the Internet

The new communicative designs are conceived to expand human possibilities. They start from specific aims and require the support of technology to achieve them. Methodologically, there is an internal process of development, which requires the external support of information and communication technologies (ICT) (Arrojo 2015). Thus, their configuration is not in nature or in the very society in which they are developed. These new designs are in the field of the artificial, which is in the sphere of the human-made, where conceived by humans materialize (Simon 1996). In this regard, creative designs and technological developments have a bidirectional dynamic, as they constantly feed back into each other. On the one hand, the development of ICT through the Internet gives support to new communicative designs and, on the other, the advance of the Internet conditions the designs themselves of the communicative phenomena and their possibilities of change (Arrojo 2017, pp. 425–448, especially p. 429).

To the extent that they are part of applied science, the new communicative designs on the Internet are focused on the solution of concrete problems. They are within a well-defined sphere of problems and with clear aims. When they seek optimization,[3] they use creativity in the treatment of information: to be able to inform more or better, to do it in a more accessible way, etc. But this would not be possible without technological innovation. These designs can be studied in terms of the sciences of the artificial. They include two main components: (a) they try to enhance the possibilities of human communication through a scientific design, looking to make a professional practice progressively more scientifically based (for further discussion of the distinction between 'design' and 'scientific design' see Niiniluoto 1993, especially pp. 9–11; and Gonzalez 2007) and (b) they need to be supported by a technological development that makes the scientific design viable (on the interaction between scientific creativity and technological innovation, cf. Gonzalez 2013a).

Based on these two main components, there is a constant interrelation between the scientific creativity of communicative designs and technological innovation related to the Internet as a network. The focus of the scientific creativity is mainly related to the Web (Gonzalez 2020a), where most of the new designs that enhance human communicative possibilities are both in longitudinal and transversal terms (Arrojo 2017, p. 427).

10.1.1.1 Designs with Longitudinal Novelty

Novelty in the longitudinal sense involves extending horizontally. Thus, this novelty makes it possible to expand the possibilities of something that already existed. This is the case when already known communication systems are improved in some directions. Designs with longitudinal novelty modify what is already known, both structurally (aims, processes and results) and dynamically (changes over time). In this regard, they make it possible to extend the uses, forms, supports, contents, etc., of a type of communication that was already being produced (Arrojo 2017).

The longitudinal novelty has allowed, for example, the hegemonic media in the analogue field (press, radio and television) to enter the digital sphere of the Internet (in the broad sense). This has happened with a clear evolution of their designs. These are now cyberspace media,[4] which can be placed on the 'third environment' (Echeverría 1999, 2006). Nowadays, they have designs with much more sophisticated aims. They can carry out processes and achieve results that were previously almost unimaginable.

There have been structural and dynamic changes in communicative designs due to the emergence of the press in the digital world. At first, newspapers saw the possibility of reaching new audiences, in different media, with multiplatform communicative content and promoting interactivity with the reader. This introduced new processes in the work routines of professionals, who were forced to incorporate different technological advances (Salaverria 2019).

The result obtained with new designs and technological improvements was the development of new formats, especially in terms of time. The consequence was to change the dominant concept of 'daily newspaper' (Arrojo 2017, p. 440). With new designs and the use of technological improvements, it was possible to go far beyond the fact that information was only updated once a day. Information is now a flow, constantly updated. Furthermore, it can remain available on different platforms.

In addition to the variations introduced at the temporal level, there were other changes with new designs: (a) the new designs multiply and diversify the information, which is now offered in many ways (hypertext, interactive infographics, native content for social networks, audio, video, etc.); (b) these variations have increased the possibilities of interacting with the public, which can happen now through web pages, advertisements, social networks, applications, etc.; and (c) the new elements have facilitated the access to content in a number of ways.

Moreover, changes made with digital press designs have also led to changes in the communicative designs of paper press. As a consequence of the variations, two visible phenomena can be noticed: (i) the paper edition is disappearing because of economic costs and a clear decrease in the number of readers in this format, and (ii) the paper edition has moved the focus towards some contents: editorial, background or explanatory information, meanwhile the digital version available through the Internet is focused on current affairs information.

Important changes have happened with the radio. The difference with the press is that radio has always been characterized by immediacy instead of being something you had to wait for the next day. The main difference between digital radio and analogue radio is that the content can also last over time. This includes new digital formats such as recordings, known as 'podcasts.' Digital radio can be followed through various media (conventional radios, cell phones, computers, etc.) and over distances that are greater than previously, since digital radio can be followed from other countries or other continents. Thus, the new processes have extended the capabilities (contents, space and time) of the radio in a longitudinal way.

The audiovisual change in digital television is also based on the new communicative possibilities opened up by the Internet and on the use of the Web as the main route. The modification of the communicative designs in recent decades changed the aims of television. The conventional television – used to be a linear television – is being relegated to the broadcasting of live events. But the new communicative designs open new possibilities like the social conversation. These events still require broadcasting on a conventional channel to have a wide audience in a specific territory.

There is a broad audiovisual offer based on the new communicative designs and the developments in the network of networks. There are now new technological instruments (cable, Wi-Fi, etc.) and a universalization of digital connection from any place and at any time. The consequences are clear: (1) the new designs increase the possibilities of interacting with the public; (2) the new designs multiply the types of information, as well as the different and complementary content that can be offered; (3) the designs expand the temporal concept of the content; and (4) the new designs multiply the audiovisual offer as a whole and facilitate access to the contents.

(1) Digital designs and the new technological developments have led to audience interaction. Now, it is possible to monitor the behavior of

individuals and the way in which they access content. Furthermore, audiovisual media can talk directly to the users and listen to their conversations (among themselves and with the content providers). This can be done through the sub-platforms that have been developed through the Web, such as social networks, and the practical applications (apps).[5]

(2) These sub-platforms can also provide access to content different from that offered on the main platform (the TV channel), giving complementary information. Thus, through social networks of the Web, there are new types of television, some of them known as 'social television.' (3) These sub-platforms have expanded the possibilities of accessing content beyond previous limits. (4) Social networks are also becoming producers of professional audiovisual content, organizing platforms where users can distribute their own content (such as YouTube, Instagram or Snapchat).

Through creativity in communicative designs and the use of technological innovation, we now also have over-the-top television (OTT), such as Netflix, HBO, Amazon, Movistar+, etc. Creativity and innovation have facilitated the emergence of on-demand television using the technological infrastructure of the Internet. These OTT media services allow access to their content at any time of the day; the user does not need to be subject to the programming set by a channel. In addition, the user does not need to have television channels.

10.1.1.2 Designs with Transversal Novelty

Transversal novelty makes it possible to move the field of what is already known in a vertical sense, so that it leads to net creativity. This novelty affects the structural level – aims, processes and results – as well as the dynamic level (change over time). It enables tasks or functions that did not exist before, such as the application of binary code, which allowed the transition from analogue to digital processing of information. This move made previously almost unimaginable tasks and processes possible (Arrojo 2017, p. 441).

This type of transversal novelty is often the scientification of a professional pattern or the result of a new design. In this regard, following scientific creativity and technological innovation, new forms and models of information on the Internet have been possible. Thus, e-mail, blogs, social networks, information search engines and apps have become reality. These are five examples of transversal novelty in communication related to the network of networks:

(a) The emergence of e-mail allows almost immediate sharing of textual information, enriched with videos or attached documents. This has made it possible to obtain communicative results unprecedented in any other type of information in the analogue environment. This novelty, which was later expanded with the webmail, is carried out with the fluidity of a telephone conversation, or any other type of oral information (Floridi 1999, p. 72).

(b) Blogs have provided a new format for communicability, insofar as they give the possibility for individuals or groups to become producers or disseminators of expert knowledge. This format also has a social dimension when it creates communities or networks around different topics. This includes a second circle, the dissemination that the hegemonic media could make of them (as has happened with blogs with a medical or political component).

(c) Social networks achieve new communicative goals with a wider radius of action than blogs. They do so according to an increasing level of specialization, even though they can share common elements. Social networks have changed over the time, but they have their own characteristics in terms of their communicative aims, such as the target audience they address and the durability of the message or the type of information they transmit. The aims condition the processes of social networks and modulate the results sought through this way of communication.

(d) Information search engines offer an important transversal novelty, since they classify and prioritize the information circulating on the Web. They do so with a variety of content: text, images, videos, maps, news, etc. For users of the network of networks, these engines have become essential elements in the 'third environment.' However, the criteria they follow when prioritizing information are not particularly clear and explicit.

(e) Apps are limited practical applications that multiply the possibilities of the network of networks. The creativity of the apps' designs makes it possible to achieve new aims for individuals, groups and organizations. They make users' time more profitable and offer different types of information in a more limited and convenient space for the user. Furthermore, they involve greater mobility and possibilities for interpersonal communication.

10.1.2 Historicity in Communicative Phenomena on the Network and Ontological Levels (Micro, Meso and Macro)

When aspects of digital communication are developed in the field of the sciences of the artificial, there is a constant interrelation between the scientific creativity of new designs and the technological innovation of the support. This interaction gives rise to new designs, which lead to new developments in the sphere of the Internet – in the broad sense – where Artificial Intelligence is involved (the structural configuration of the Internet involves scientific creativity and technological innovation, which are dynamically interrelated. For further information, see Gonzalez

(2017), especially, pp. 401–403). There is thus a continuous flow of new designs, which require new technological innovations, which in turn make it possible to configure new communicative designs. From a structural viewpoint, these designs ensure the progress of the network of networks as well as of the scientific disciplines that use the Internet – through its layers – as a support for their development. From a dynamic perspective, the scientific and technological aspects of the Internet – in the broad sense – are intertwined.

Besides the scientific side and the technological facet, the social dimension of the network of networks should be taken into account (Gonzalez 2020b). The new communicative designs, through the support of technological innovation, take place within a society – individuals, organizations, companies, etc. – that has a given socio-political, economic and legislative framework. This society either adopts the new communicative uses (the new platforms and the new content) or rejects them, while at the same time demanding new services. *De facto*, there is a constant interrelationship between new scientific designs – the internal component of the Internet – and external factors, which are driven by the society in which they develop.

To understand the new communicative designs, attention is required on both the structural level – the aims, processes, and results – and the dynamic level, which is the change over time. This involves historicity in the constant interaction between the scientific side, the technological facet, and the social dimension of the network of networks. In this regard, the approaches on complexity presented by Nicholas Rescher and Herbert Simon are insufficient to grasp the dynamics of historicity.

Rescher offers a general approach to structural complexity. He does it from the epistemological and ontological point of view, but he does not deal, strictly speaking, with the sciences of the artificial. Regarding dynamic complexity, it only appears collaterally in his position, when the functional complexity can come to have a dynamic vein (Rescher 1988, p. 9. In the case of communication sciences, see Gonzalez and Arrojo 2019). In the case of Herbert Simon, when he analyzes the complexity of the sciences of the artificial, the focus is on the structural complexity. The dynamic complexity only appears insofar as he accepts the idea of evolution in the artificial systems (Simon 1996; and Simon 2002).

When the attention goes to dynamic complexity, three concepts should be considered: processes, evolution and historicity (Gonzalez 2013b; and Gonzalez 2018b). The first, which is dealt with by Rescher, is applicable to new communicative designs on the Internet (in the broad sense). Thus, there are 'processes' in the sense of 'productive processes,' which are those that can be identified with something operational (e.g., the steps that must be followed to design a social media campaign that leads to the expected results), and in terms of 'state transformation processes' (e.g., the 0 and 1 coding of content, which allows for the transition from analogue state to digital expression) (Rescher 1995, pp. 60–62).

'Evolution' can also be seen in new communicative designs, as is often the case in the dynamics of complex systems. Normally this dynamic of evolution is associated with living things and adaptation to the environment (Crutchfield and Van Nimwegen 1995; Kryssanov et al. 2005; and Gonzalez 2008). But this concept fails to capture the degree of complexity of today's communicative designs via the network of networks. These new designs present an increase in the artificial character of communication and a high level of novelty. Thus, an increase in dynamic complexity cannot be explained only from the perspective of processes and evolution.

Therefore, Simon and Rescher's approaches are insufficient to analyze the dynamic complexity of communicative designs in the Internet domain. This type of complexity is much better understood if we use the concept of 'historicity,' proposed by Wenceslao J. Gonzalez, to characterize the dynamics of the sciences of design (Gonzalez 2013b, especially pp. 304–307; and Gonzalez 2011). He points out that the two categories mentioned above – 'process' and 'evolution' – are insufficient to reflect scientific change and need to be complemented by the category of 'historicity' (Gonzalez 2011).

10.1.2.1 Historicity to Explain the Revolution of Communicative Designs

There has been a revolution in the field of communication through the spread of the Internet (Floridi 1995). The idea of historicity can explain this emergence of new communicative realities much better than the notions of process or evolution. De facto, the change brought about by the network of networks is an authentic revolution in the history of communication. This can be recognized by the characteristics of 'revolution:' (a) it is a radical change with respect to the previous situation and (b) it is change that has not been prepared in any relevant way and, in this regard, it is an unforeseen phenomenon. This is the case especially when it is a transversal novelty. Thus, it is a phenomenon that cannot be reduced to mere terms of evolution.

Since the use of the possibilities of the Internet – mainly through the Web – new communicative designs have emerged. Some of them involve longitudinal novelty and others have transversal novelty. The constant interaction between scientific creativity and technological innovation led to novelty. There is also an impact on the social dimension of the network of networks. Furthermore, there is complexity both on the internal sphere, based on science and technology interaction, and on the external ground, due to the social dimension. This complexity is initially structural and then dynamic. On the internal sphere of the Internet, historicity is related to the variations in the set of elements that configure the sciences of communication (language, structure, knowledge, methods, activities, aims and values) (Gonzalez 2005, especially pp. 5–6) as it develops through the network. On the external ground, historicity also makes it possible to explain the threefold change that takes place: in the agents involved, in the progress of communicative research and in the communicative reality itself, which is also subject to change.

Regarding science, there are three main levels of historicity: in science as a human activity, in the agents and in the reality under study. The focus can be placed on the agents. Thus, three complementary aspects are involved here: (1) the agents live in a specific historical environment, which varies over time; (2) these agents relate to others, which entails variations in the centers of research; and (3) the agents interact with a changing communicative reality, which modulates their work (these aspects are relevant to understand the external facet of the Internet, cf. Gonzalez 2020a).

The historicity of the dynamic complexity due to the Internet has appreciably modified the types of communicative relations, both in the artificial facet – the new Internet supports – and in the social dimension of an interpersonal nature (Scolari 2018). The very reality of the network of networks is constantly changing, influencing both internal and external communicative processes. These Internet developments have led to an increase in the artificial character of communication, which starts with the designs and receives the social echo. This leads to structural complexity and dynamic complexity. This is often not a mere evolution but rather a revolution within, which can be explained in terms of historicity (Gonzalez 2013b, especially pp. 304–307; and Gonzalez 2011).

10.1.2.2 *Micro, Meso and Macro Levels at the Structural and Dynamic Levels*

Concerning structural and dynamic complexity of the new communicative designs, there are several ontological levels. These are the micro, meso and macro levels, which should be taken into account to have the whole picture. Thus, the ontological novelty might be in any of these levels. There is, commonly, a novelty at the ontological level when there is a change in the network of networks, which is where digital communications take place. This change has an impact on the dynamics. In this regard, the search for new improvements in design leads to the development and implementation of new processes. These changes are usually influenced by society, which uses the Web and fosters new goals.

When it comes to communicative designs with transversal novelty, as in the case of e-mail, there are variations according to levels. Thus, the communication at the micro level is between a few individuals. Meso is when there is communication between employees of a medium-sized company, usually regional or national. Macro is when there is communication between companies the size of Apple or Samsung or international organizations. Meanwhile, there might be blogs conceived for small communities, such as a classroom (micro), for a company with a nationwide presence (meso) or for a site with a purely international reach (macro).

In the case of social networks, the ontological levels have an impact on structural complexity, mainly at the time of their configuration, but also on dynamic complexity, which is the search for new aims based on the interaction with the public that uses them. Thus, there are communication strategies at micro level, with direct messages between a few users, at meso level, when they allow the creation of large groups of friends, or at macro level, when they want to reach as many people as possible (such as voters in a general election). In addition to this type of communicative design, there are social networks that are conceived with the vocation of addressing more restricted communities, either by subject matter or sector of interest, such as academia, or by sector, such as audiovisual workers, or by the age of the target audience.

Apps can also have a micro level of communication when they want to foster a direct relationship with the individual, encouraging them to carry out concrete actions. They can be at the meso level when they develop communicative actions focused on a specific community, such as apps for finding accommodation or transport services in specific geographic areas, apps for television entertainment formats, which are broadcasted in specific territories and are confined to specific geographic areas. Apps can be macro when they seek to reach as many people as possible. The communication designs for each of these levels will be different, although they will have to coexist with each other, and the passage of time leads to a review of aims, processes and results.

Following these ontological levels and the previous reflections, we have a philosophico-methodological framework for the analysis of the communicative design of a social network such as Snapchat (which also has other aspects, cf. Redacción Tecnosfera 2017). It is the approach to sciences of communication as sciences of the artificial. The combination of epistemological, methodological and ontological aspects considered here are relevant for this recent case. Snapchat illustrates the revolution that the Internet has brought about in communication. The study addresses both the internal sphere and the external ground.

10.2 A New Communicative Design on the Internet: Snapchat

Using a new communicative design, the social network Snapchat expresses a clear interaction between scientific creativity and technological innovation. Its design is an alternative to existing social networks and complements them. It significantly increases the artificial part of communication and opens a gap in the audience of social networks. As a consequence, Snapchat conditions the position of the other established players, leading to changes in other social networks When Snapchat burst into the market, there were other well-established social networks with very clear objectives.[6]

Twitter, for example, sought to encourage interactivity and speed of conversation by placing a limit on the number of characters in posts. Meanwhile, Facebook sought to be a general network, where the message would last, in order to build an online social reputation. LinkedIn, on the other hand, was born as a network with a more sectorial character, as it was focused on the workplace. Its goal was the active search for employment or the promotion of labor relations.

Snapchat's uniqueness lays in three features: (a) the social network allowed instant messaging for the first time, as the technological resources were already integrated into the application's own mobile device, (b) the messages it transmitted were ephemeral, as they were self-destructed after a period of time and (c) it safeguarded the identity of its users. Thus, this social network was the first to use technology to make posts ephemeral. This represented a real revolution in the field of innovative social networking. With its new communicative design, Snapchat attempted to provide a solution to a specific problem: the durability of messages on social networks – because it can condition the social reputation of individuals in the future. This approach was revolutionary at the time. This quickly led to thousands of people starting to use it. Snapchat also got other social networks to adopt various designs. In this sense, it was the origin of a new communication model in the digital environment.

10.2.1 The Communicative Structure of Snapchat: Epistemic Complexity and Ontological Complexity

The new communication phenomena via the Internet are based on designs with clear objectives. These are three in Snapchat: to be immediate, to send ephemeral messages and to protect the identity of its users. Thus, it seeks creativity in the treatment of information, in order to reach new communicative heights. It is similar to oral conversation, insofar as it leaves no trace of 'conversations' via the Internet between individuals or between groups. Unlike what happened in the rest of the existing social networks at the time, the content disappears when it is seen or listened to.[7]

Analyzing this from the point of view of structural complexity, which can be epistemic or ontological, there is epistemic complexity here in terms of formulation (Rescher 1988, p. 9). In this sense, it tries to simulate the occasional conversation between friends using a social network. Snapchat's communicative design uses a syntactic basis. These signs – incorrectly called 'symbols'[8] – require an interpretation to have a semantic content. The programs used in the case of Snapchat have sought to enhance the access, processing, retrieval, dissemination, and elimination of this information. This new communicative design can perform the task when the following artificial devices are available: (i) a technological communication infrastructure with access to a wireless connection (Wi-Fi); (ii) the use of recent versions of mobile phones (smartphones); (iii) a specific practical application (an app); (iv) a geolocation system; (v) an object or person identification system; (vi) an augmented reality system and filters; and (vii) an encryption system that protects privacy, which allows the storage and subsequent destruction of data.

The elements that Snapchat uses for communication can be structured in two stages or ranks: the first is photographs and videos, which guarantee a very simple interaction, and the second, which can be added to the previous one, has the following elements: (a) very limited text messages, (b) finger drawings (doodles), (c) emoticons, (d) gifs, (e) stickers,[9] (f) geofilters and (g) augmented reality masks to take selfies.[10]

These elements are accessed through the device itself, installed in a mobile phone application. This allows scientific designs to continue to advance as well. It also allows for greater operability. This leads to the analysis of three different levels of formulation: (1) the type of communicative content that is produced, (2) the type of communication that can take place in terms of duration and (3) the identification and registration of users.

The different levels of communication possible are established through the application itself, installed on the mobile phone. Thus, the user can communicate in a particular way with a friend, by sending a message, through chat, or by means of a novel development of this application, called 'stories,' to which all users can have access. If a 'story' is sent, the duration of the story is 24 hours, and the recipient can view it an unlimited number of times during that time.

(1) For immediate communication, the content sent through this social network is essentially static or moving images, which may be enriched. In the case of private messages or chat messages, when the sender sends the content (photo or video), this content automatically disappears from the sender's device.

(2) The aim is to send ephemeral messages, so you can choose the time it is available: you choose how long the message can stay on the recipient's device when they open it (you can choose from 1 to 10 seconds).[11] Recipients can capture that content, but when this happens, the sender will receive a notification.[12]

(3) Users' identities are protected. Thus, to register on Snapchat, only a phone number and an alias are required. This social network does not aim to create a user 'profile' as such, with personal data such as date of birth, tastes, academic or professional background, etc. The aim is to protect the user's privacy and not to display it for others to 'discover.' In fact, the only way to search for a user is by knowing their pseudonym on the social network.[13] Young people have valued the fact that they can communicate with their close friends for real, that only those who know them can find them.

These current functions – and future ones as new information demands arise – make this social network very dynamic. It thus has higher levels of complexity than other social networks, in terms of the formulation of the aims and the characteristics of the processes to achieve the desired results.

10.2.2 Communicative Dynamics of Snapchat and Levels in Communication

Alongside the structural complexity – which here has focused on the formulation – of Snapchat's new communicative design, there is the dynamic complexity. This is based on the internal and external factors of the change operated by this social network. Initially, the sources of change are two: (I) the agents responsible for this social network themselves, who continually seek new advances in the designs, leading to the development and implementation of new processes to anticipate or meet new user demands and (II) the social receptivity of the changes when it comes to adopting or not the new designs, which modulates the permanence of the designs or the configuration of new ones.

There is a twofold communicative dynamic, which generates complexity in Snapchat. First, there is the complexity that is structural in origin and is open to change. This communicative dynamic is *ad intra*, as it arises from the interaction between scientific creativity and technological innovation. Secondly, there is the complexity that originates in social acceptance. This is the dynamic complexity *ad extra*, which appears when the network introduces a change, a distinctive or differentiating feature, and there is a social response, which is then internalized by Snapchat. Thus, users show a degree of acceptance and, in turn, other social networks or other communicative agents respond to the change, either by adopting it or by relocating to this communicative environment.

In 2011, when Snapchat was created, it was an instant messaging-only application. These were messages that disappeared as soon as they were seen. This differentiating feature was not understood at first by the communication industry. It went in the opposite direction to one of the main objectives of the other social networks, which is to endure over time and create a digital identity for users, precisely through their posts on these networks.

Due to the apparent complexity of its design, Snapchat's use was a deterrent for people that were unfamiliar with the digital environment.[14] Thus, its target audience were users who came of age at the turn of the century (millennials). This led to a clear segmentation by age. Initially, as they felt that their privacy was being preserved, teenagers exchanged content, both images and videos, with a high sexual content. More adult profiles (parents, relatives, etc.) were on other social networks, but not on Snapchat. But the users soon discovered the potential of this network (a) as a means of direct communication with close friends and (b) as a network that preserved users' privacy.

Among young people, sharing photos and videos is a widespread act of communication (Kofoed and Larsen 2016). However, not feeling monitored and the immediate disappearance of messages led them to develop more spontaneous content, within an environment of shared intimacy (Obake and Ito 2006). In 2013, Snapchat added the functionality 'stories' (my story). With this new functionality, users were able to instantly post messages they had created themselves. They were posted on a public board, accessible to all users of the network, not just their friends. These messages disappeared after 24 hours.

The network went a step further when it offered this new macro communication platform for a new type of stories. In them, users could post multiple images or videos, in an order chosen by themselves. In this way, they made this narrative into small stories that were only active for 24 hours. From then on, it offered three levels of communication: (i) content focused on one person or on the followers selected within the profile; (ii) content aimed at all the profile's followers; and (iii) content aimed at all the platform's users. This is when the app's popularity soared.[15] In 2013, according to data provided by the company itself, 14,000 million photos and videos were shared daily, and 500 million 'stories' were played per day (Lacort 2018).

In 2015, the network launched 'discover' (Ortiz 2015). This is a carousel of stories produced by editorial teams from different media. At that time, Snapchat had already managed to attract an audience that was highly segmented by age: they felt comfortable with the application and produced a different type of content that was spontaneous and natural. Editorial media companies – CNN, Buzzfeed, National Geographic, etc. – and brands then saw an opportunity to approach this audience through the platform. For Snapchat, this is a new business model, which allows them to sell digital spaces or channels that are adapted to the profile of their users.

In 'discover' you can see the short messages or snaps (10-second videos) that the media create. This is an innovative way of integrating editorial media into this social network. Thus, in addition to being a social network, it becomes a place for discovering news. This new facet was accentuated with the creation of 'live stories' – a space within 'discover,' which, through short messages (snaps), allows live coverage of a massive event, such as the Oscar ceremony or the Super Bowl. These videos are paid for by the media that create them. This also allows users to upload snaps of a live event or a city chosen by Snapchat and make it visible to the network's millions of users for a day (Verstraete 2016). This new functionality contributed to the international expansion of Snapchat, which then became popular in Europe and Latin America.

This integration into a social network of other media companies represents an example of longitudinal novelty in two ways: (1) the media find a new platform through which to reach their audience and (2) for the first time, a social network opens up a defined space for third-party editorial content.

Snapchat then – in 2015 – introduced another technological development with new functionalities: filters that made it possible to create illustrations and animations on users' faces. This was done for several reasons: (a) to further differentiate itself from other social networks, (b) to intensify the feeling of belonging to a group, and (c) to generate active engagement with the platform and the way of expressing oneself on it. This functionality was a real success. It ensured that the application continued to grow in the three major territories where it was already present: the United States, Europe and Latin America.

At the same time, competing social networks were forced to incorporate these functionalities into their platforms. In 2013, Mark Zuckerberg tried to buy Snapchat. Faced with the refusal of its CEO, in 2016, he began to adapt some of the innovations of this social network to his own. Thus, in 2016, Instagram incorporated filters with which to modify photographs and videos. It then launched Instagram stories.[16] The difference with Snapchat is that it allows the uploaded photographs or videos to be archived. This new Instagram feature was well received by the audience. It quickly led to an increase in the number of teenagers and young people joining Instagram.

It was then – in mid-2016 – when Snapchat launched *memories*, a feature that also allows archiving of photos and videos uploaded to stories. This social network breaks with the strategy of the pre-eminence of instantaneousness and transience, in order to come up with another longitudinal novelty, which responds to the advances of another social network and allows it to maintain its users (Galeano 2016). There is an internal dynamic here, since creativity in the designs and the use of appropriate technology allowed new functionalities to be incorporated. But there is also an external origin and it involves structural changes in the network. Thus, in mid-2016, the group conversation and video call function were incorporated into Snapchat, as well as the possibility of making voice notes and sending them through the conversation or chat mechanism (Mejía 2017), to offer its users new ways of communicating through its platform. Two years later, in 2018, this group conversation through video calling allowed you to chat with 16 friends at the same time.

As a result of this dynamic, which began in 2011, Snapchat went public in 2017, dragged by market conditions. It then made a foray into the technology market,[17] in order to continue to differentiate itself from its competitors and stem the flow of users who were abandoning its services. It launched Spectacles, a pair of glasses that could be used to take photos or record videos,[18] only compatible with its social network (Vega 2017). But the sales results were not as expected (Vega 2017).

By expanding the structural complexity, it obtained an adverse result. It was in 2018 when it made a redesign of its application, which generated a backlash from its followers. The main novelty of the redesign was the division of content produced by friends, located on the left, and by institutions, brands and influencers, located on the right. Some users complained that usability was now more confusing. The company echoed these complaints. It re-implemented snaps and chats in chronological order. It put friends' stories back on the right side of the app, although they remained separate from branded content. For the first time since its inception, towards the end of 2017 and during 2018, it recorded a drop in the number of daily users.[19]

10.2.3 Role of Agents in Snapchat: Individuals, Groups and Companies

For a communicative design to be fit for purpose, two factors are relevant: the ontological level in the design goals and the interaction by the network users. First, there is how this design fits into the different ontological levels in the social environment. These are three: (i) the micro level, when the novelty of Snapchat for the relationship between two people is addressed, (ii) the meso level, when the novelty introduced by this new design covers wider communities, such as groups of friends or larger collectives, and (iii) the macro sphere, when the designs reach the whole of the application open to the whole world, which means many millions of users. Secondly, there is the use that agents make of this communicative design. Because users are not passive, so their response can bring creativity. Consequently, they can change the processes followed up to that point, leading to new forms of technological innovation or even the redesign of the whole platform.

Snapchat has shown a history of seeking novelty in its interaction with its users, rather than looking for an internal evolving in its dynamics from the beginning. Thus, it has come to cover the different spheres of communication according to micro, meso and macro levels in the objectives, which can be rethought from the perspective of economics (for further information on micro, meso, and macro taxonomy applied to the economic environment see cf. Dopfer 2004; and Lijenström and Svedin 2005). It is also possible to appreciate different levels from the users' angle, since the behavior of the agents can be situated in the mere sphere of strictly private communication (micro), or they can reach a greater relationship of a social nature (meso) or, potentially, they are faced with a greater interaction when it is commercial or business within a broader environment (macro).

(i) At the micro level, young people's use of photos or videos highlights their interest in everyday life, entertainment, etc. They need to express their feelings and socialization among peers. At this level, Snapchat functions as a direct messaging system, like other social networks. It is individual-to-individual communication. The fundamental difference is that it is ephemeral messaging.

(ii) Meanwhile, at the meso level, Snapchat also differs from other social networks by emphasizing trust. Snapchat encourages the idea of connecting friends who are real friends – the inner circle – and they are provided with the username that is used, even if they are somewhat broad collectives. It seeks to embrace a safe environment, where they can feel free and safe from prying eyes (Rubio-Romero and Perlado 2017). Thus, Snapchat is the only practical application that does not allow videos to be shared on websites, blogs or other digital spaces (Antevenio 2016).

(iii) In the macro sphere, insofar as it includes media companies that can reach many thousands of people, the operation of Snapchat was a real challenge. Brands and media outlets had to develop a new type of format to coexist with the 'stories' uploaded by users. This is the case of the vertical video format, of short duration (10–20 seconds), which only last 24 hours. The aim is to capture the attention of young people, based on naturalness, and where the editorial focus is to cover points of view that are not covered in other ways.

With Snapchat, media companies are looking for closeness: the more human side and telling stories live, paying attention to what is happening behind the scenes. This has led to a novelty in advertising: coupons exclusively for its users, taking advantage of the fact that content is deleted within 24 hours. It is a way to generate active engagement and avoid oversupply.[20]

10.3 Novelty Factors Provided by This New Social Network

When we compared Snapchat to the social networks that existed at the time, there are four differentiating elements that Snapchat originally brought to the table: (1) the management of ephemerality, (2) the different approach to privacy, (3) the use of image filters to generate active engagement with the user and (4) a novel way of allowing the media and brands to enter the platform, with a different advertising and business model. In them, a new communicative design is palpable, which has marked a before and after in the history of communication on the Internet.

10.3.1 *Uniqueness of Content: Prevalence of Ephemerality*

Snapchat's main feature is that it emulates real conversations. This brought it clear success at the time of its launch (Kamleitner 2018; and Redaccion Canela PR 2017). From an epistemological point of view, what prevails is the private – the intimate sphere. From the ontological perspective, the ephemeral prevails. Thus, it revolves around something that flows quickly in time. This social network attends to the very moment in which things happen. It does this through the photos and videos that can be captured with the application itself.

It has a 'dialogic' structure, in that it seeks to recreate the action of having a direct conversation in several ways: (a) the type of communication is based on the spontaneity of the moment and the exclusivity of choosing an interlocutor and (b) the message, once seen, disappears within a maximum period of 10 seconds. It flows, somewhat like words, and is only part of the memory of those who are on the subject. This unique moment of viewing that content, like words, can never be shared again. This new type of communication was a transversal novelty, due to its somewhat revolutionary character.

Instead of enduring the event, the aim is the transience of the message. It is, thus, the first time that a completely different use of the resources of the Web appears: the Internet based communication platforms. Up until Snapchat, social networks located on the Web were mainly focused on what is stored and archived: Moments were divided up, placed on a platform, quantified and classified according to the sections of each profile, etc.

10.3.2 *Volitional Aspects and Evaluative Rationality*

The desire to do something with ephemeral content makes a difference: the user does not need to revise or perfect the content to be emitted. Thus, the ends sought by evaluative rationality are, in principle, all transitory. In addition to spontaneity, there is more room for feelings and emotions, as in the case of conversation via spoken language (Skierkowski and Wood 2012). The desired goal is the conversation itself. This is what users valued and what distinguished this social network. The other social networks provide much more data about the identity of the individual in the ordinary (offline) world: the profile picture, the community of friends, the friend suggestions, the indication of what they like or what they do, etc. All of these provoke a sense of control or surveillance, which Snapchat's network does not have.

By seeking out everyday reality, Snapchat encourages sentiment and the expression of emotions (Kamleitner 2018). This, in turn, encourages users to send many more messages and to always be connected. In this regard, Nathan Jugerson – researcher and sociologist – explains that the aim of this social network is that citizens are not classified (Jugerson, 2013). Consequently, the profiles themselves do not structure identity, in a more or less restrictive way.[21] Jugerson insists that only by establishing a structure in which content has an expiry date – and, therefore, cannot be used to pigeonhole us or cost us a job – can we truly feel free on the Web. This is the approach on which Snapchat is built.

It enables Snapchat to show itself in a more natural way. Users of this network move away from constructing an ideal persona in the digital environment to show who they are in real time. This is in line with the rise of live video – the authentic and the real versus the posturing and the perfect.[22] Snapchat does not encourage the number of friends, but rewards activity. The most active users are rewarded with trophies to encourage them to use the application even more. It differs from Facebook, where the friend request is in fact a request to establish a communication network.

Snapchat's type of communication rewards participation, and ephemerality has a number of consequences. First, the fear of missing out,[23] which leads users to keep an eye on their mobile phones. It involves always staying connected, to check if their friends, brands or favorite celebrities have posted something new. Second, authenticity in the fact that the user expects to find unedited videos, authentic images or spontaneous comments in real time. This is an opportunity for advertisers, because it creates a stronger emotional and trusting bond. Third, active engagement, as users can create or influence content with filters, stickers, drawings, etc. Fourth, it involves another type of 'programming,' due to its ephemerality. What is exciting about this is that it is liquid and unexpected programming.

10.4 Final Considerations

In retrospect, Snapchat's novelty is clear, in that it became a very attractive social network for young audiences. This was due to six main reasons: (i) it is free; (ii) its navigability was not simple for more adult people; (iii) it was possible to hide real identities; (iv) images had a lifespan of seconds before self-destructing (Sulbaran 2016); (v) it leaves a field open to creativity; and (vi) the type of communication it encouraged was relational.

The underlying novelty comes from the design, which seeks similarity with informal expressions of entertainment and relationship. It is about being with friends – not to have deep conversations but simply to entertain and socialize. This playfulness means that Snapchat's innovations cater to the flow of everyday life: snapshots and live videos, spontaneous and lighthearted. They are like 'televised' lives, which emphasize the importance of sharing something with someone (Kofoed and Larsen 2016) and doing so at the very moment it happens. It is about sharing, either in a private way (the individual or micro sphere) or in larger environments (the social or macro sphere).

Along with the novelty of design is process innovation. Thus, Snapchat highlights the ephemeral rather than making – as it happens in other social networks – a cyber-identity base, which is used as a repository of life memories for social projection (Rubio-Romero and Perlado 2017).

These processes, which have produced novel results in terms of social interaction, especially for young people, have imitators. This is the case of Facebook's competition (Lacort 2018).

The technological basis that has enabled Snapchat's innovation is easily replicable. Other social networks or messaging systems (e.g., Instagram, Facebook or WhatsApp) have joined the trend of ephemeral content and entertainment (filters, etc.). Suddenly, ephemeral content and immediate engagement results, which made Snapchat different, are ubiquitous. Snapchat has lost that exclusivity. There has also been a shift of users to Instagram, where the cyber community is larger and, consequently, so is the impact of posts.

Notes

1 This paper has been written within the framework of research project PID2020-119170RB-I00 supported by the Spanish Ministry of Science and Innovation (AEI).
2 Some of these sciences are Network science, Web science, Internet science, and data science. See Gonzalez (2018a, p. 77); and Tiropanis et al. (2015).
3 Optimization is a different concept from maximization and certainly different from satisficing.
4 The longitudinal novelty has allowed, for example, the hegemonic media in the analogue field (press, radio and television) to enter the digital sphere of the Internet (in the broad sense). This has happened with a clear evolution of their designs. These are now cyberspace media.
5 Practical applications (apps) belong to the third major layer of the network of networks, which also includes the cloud computing and the mobile Internet. Cf. Gonzalez (2022).
6 Snapchat is a mobile application developed in 2010 by three Stanford University students: Evan Spiegel, Bobby Murphy and Reggie Brown, as a final year project. It was initially called Picaboo, a play on the word peek-a-boo, a children's expression related to the fright of a ghost. It is part of the company Snap Inc. The company's motto is 'Live in the moment, share and express yourself.' See Garcia (2017) and Soto (2017).
7 Messages don't need to be viewed at the time they are sent, although they do have an expiry date and are deleted after that time if they are not opened. There are other types of content, such as 'stories,' which have an expiry date of 24 hours. However, what this network seeks is real-time communication.
8 Genuine symbols are polysemantic, as is the case with words like 'crown,' which can be something material, something institutional or the person representing the office at the time.
9 Snapchat has a large number of stickers, in addition to traditional emojis, that streamline communication between users. It is associated with Bitmoji, a company that allows the creation of personalized stickers that interact with those of other users. See Cuartas (2016).
10 In the Snapchat environment, they are also known as lenses. These are filters that recognize the user's face and allow the user to distort it, add effects, etc.
11 If a user sends a short message or snap to another user – chat, photo or video – the latter will only have the opportunity to see it twice and, if they forget to open it, the message will self-destruct within 30 days. See cf. Sulbaran (2016).

12 The system can only detect that someone has retained the content if it has been taken via a screenshot with the device itself. But it is not possible to detect if the photo or videos sent are recorded with a device other than the one on which the application is installed.

13 If users allow the app to access their phone contacts, it also identifies which of those users have their Snapchat account. But this feature is optional.

14 Snapchat's visual appearance was so minimalist that adults did not feel comfortable, as they did not understand its functionalities. See Dans (2016).

15 In 2013, Facebook offered $3 billion to Evan Spiegel in exchange for Snapchat. At that time, it was one of the most downloaded applications in the world by the so-called millennials: young people between 18 and 34 years old. See Mejía (2017).

16 First came Instagram stories in 2016. Months later, Facebook, the parent social network, also integrated stories and, at the beginning of 2017, it also included the stories service in the instant messaging network WhatsApp. Cf. García (2018).

17 By January 2019, Snapchat's shares had lost 65% of their value from when it had gone public two years earlier. See Technosphere Edition January 29th, 2019.

18 The company's CEO points out that it is a technological company and not just a social networking or media service. See Tecnosfera Edition 2018.

19 Snapchat went from 191 million users in January 2018 to 188 million in the second half of 2018. Snapchat's followers launched a petition on the Change. org website, calling for a return to the platform's original design. They got more than 1.2 million signatures. See Redacción Tecnosfera (2018) and Redacción Vida Actual (2018).

20 For example, the Taco Bell brand offered exclusive content. Its Snapchat followers were the first to discover new dishes on its menus. See Redacción Canela PR (2017).

21 Profiles structure identity in a more or less restrictive way: real name policies, lists of information about our preferences, detailed histories and current activities form a highly structured set of frames to be integrated. See, Cf. Jugerson (2013).

22 The demoscopic institute The Cocktail Analysis and Arena Media detected this trend as an emerging phenomenon in the networks, in the *VIII Wave of its Observatory on Social Networks*, carried out in 2016. See The Cocktail Analysis and Arena Media (2016).

23 Its acronym is FOMO: Fear of missing out. See Redacción Kanlli (2018).

References

Antevenio. 2016. *Comparativa del valor del vídeo en las principales redes sociales*, December 16. Available at: https://www.antevenio.com/blog/2016/12/comparativa-video-en-redes-sociales Accessed at 1.2.2019.

Arrojo, M. J. 2015. 'La complejidad como marco de estudio para las Ciencias de la Comunicación', in *Argos*, vol. 32, pp. 17–34.

Arrojo, M. J. 2017. 'Information and the Internet: An Analysis from the Perspective of the Science of the Artificial', in *Minds and Machines*, vol. 27/3, pp. 425–448.

Crutchfield, J. P. and Van Nimwegen, E. 1995. 'The Evolutionary Unfolding of Complexity', in *Proceedings of the National Academy of Sciences of the Unites States of America*, vol. 92/23, pp. 10742–10746.

Cuartas, E. 2016. '¿Cuál aplicación es mejor, Snapchat o Instagram?', in @ *ENRIQUECUARTAS*, December 1. Available at: https://www.enter.co/chips-bits/apps-software/cual-aplicacion-es-mejor-snapchat-o-instagram/ Accessed at 1.2.2019.

Dans, E. 2016. 'Snapchat y la deriva generacional', *ED*, July 5. Available at: https://www.enriquedans.com/2016/07/snapchat-y-la-deriva-generacional.html Accessed at 1.2.2019.

Dopfer, K. 2004. 'Micro-meso-macro', in *Journal of Evolutionary Economics*, vol. 14, pp. 263–279.

Echeverría, J. 1999. *Los Señores del Aire: Telépolis y el Tercer Entorno*, Barcelona: Destino.

Echeverría, J. 2006. *Cosmopólitas domésticos*, Barcelona: Anagrama.

Floridi, L. 1995. 'Internet: Which Future for Organized Knowledge, Frankenstein or Pygmalion?', in *International Journal of Human-Computer Studies*, vol. 43, pp. 261–274.

Floridi, L. 1999. 'A Revolution Called Internet', in Floridi, L. (ed.), *Philosophy and Computing. An Introduction*, London: Routledge, pp. 54–87.

Galeano, E. S. 2016. 'Snapchat Memories cambia para siempre la filosofía de la red social efímera', in *Marketing Ecommerce*, July 7. Available at: https://marketing4ecommerce.net/snapchat-memories/ Accessed at 29.1.2019.

Garcia, A. 2017. 'Snapchat: crónica de una muerte anunciada', in *La manzana mordida*, June 15. Available at: https://lamanzanamordida.net/snapchat-cronica-de-una-muerte-anunciada/ Accessed at 1.2.2019.

García, E. 2018. 'Snapchat se prepara para integrar la búsqueda visual de Amazon', in *Xatacamovil*, July 10, Available at: https://www.xatakamovil.com/aplicaciones/snapchat-se-prepara-para-integrar-busqueda-visual-amazon Accessed at 1.2.2019.

Gonzalez, W. J. 2005. 'The Philosophical Approach to Science, Technology and Society', in Gonzalez, W. J. (ed.), *Science, Technology and Society: A Philosophical Perspective*, A Coruña: Netbiblo, pp. 3–49.

Gonzalez, W. J. 2007. 'Configuración de las Ciencias de Diseño como Ciencias de lo Artificial: Papel de la Inteligencia Artificial y de la racionalidad limitada', in Gonzalez, W. J. (ed.), *Las Ciencias de Diseño: Racionalidad limitada, predicción y prescripción*, A Coruña: Netbiblo, pp. 41–69.

Gonzalez, W. J. 2008. 'Evolutionism from a Contemporary Viewpoint: The Philosophical-Methodological Approach', in Gonzalez, W. J. (ed.), *Evolutionism: Present Approaches*, A Coruña: Netbiblo, pp. 3–59.

Gonzalez, W. J. 2011. 'Conceptual Changes and Scientific Diversity: The Role of Historicity', in Gonzalez, W. J. (ed.), *Conceptual Revolutions: From Cognitive Science to Medicine*, A Coruña: Netbiblo, pp. 39–62.

Gonzalez, W. J. 2013a. 'The Roles of Scientific Creativity and Technological Innovation in the Context of Complexity of Science', in Gonzalez, W. J. (ed.), *Creativity, Innovation, and Complexity in Science*, A Coruña: Netbiblo, pp. 11–40.

Gonzalez, W. J. 2013b. 'The Sciences of Design as Sciences of Complexity: The Dynamic Trait', in Andersen, H., Dieks, D., Gonzalez, W. J., Uebel, Th. and Wheeler, G. (eds.), *New Challenges to Philosophy of Science*, Dordrecht: Springer, pp. 299–311.

Gonzalez, W. J. 2017. 'From Intelligence to Rationality of Minds and Machines in Contemporary Society: The Sciences of Design and the Rol e of Information', in *Minds and Machines*, vol. 27/3, pp. 397–424. DOI: 10.1007/s11023-017-9439-0.

Gonzalez, W. J. 2018a. 'Internet en su vertiente científica: Predicción y prescripción ante la complejidad', in *Artefactos: Revista de Estudios de la Ciencia y la Tecnología*, vol. 7/2, 2nd period, pp. 75–97.

Gonzalez, W. J. 2018b. 'Complejidad dinámica en Internet como plataforma de información y comunicación: Análisis filosófico desde la perspectiva de Ciencias de Diseño y el papel de la predicción', in *Informação e Sociedade: Estudos*, vol. 28/1, pp. 155–168.

Gonzalez, W. J. 2020a. 'The Internet at the Service of Society: Business Ethics, Rationality, and Responsibility', in *Éndoxa*, n. 46, pp. 383–412. Available at: http://revistas.uned.es/index.php/endoxa/article/view/28029/pdf Accessed at 6.7.2020).

Gonzalez, W. J. 2020b. 'La dimensión social de Internet: Análisis filosófico-metodológico desde la complejidad', in *Artefactos: Revista de Estudios de la Ciencia y la Tecnología*, vol. 9/1, 2nd period, pp. 101–129.

Gonzalez, W. J. 2022. 'Scientific Side of the Future of the Internet as a Complex System. The Role Prediction and Prescription of Applied Sciences', in Gonzalez, W. J. (ed.), *Current Trends in Philosophy of Science. A Prospective for the Near Future*, Synthese Library, Cham: Springer, pp. 103–144.

Gonzalez, W. J. and Arrojo, M. J. 2019. 'Complexity in the Sciences of the Internet and Its Relation to Communication Sciences', in *Empedocles: European Journal for the Philosophy of Communication*, vol. 10/1, pp. 15–33. DOI: 10.1386/ejpc.10.1.15_1. Available at: https://www.ingentaconnect.com/contentone/intellect/ejpc/2019/00000010/00000001/art00003 Accessed at 6.7.2019.

Jugerson, N. 2013. 'The Liquid Self', in *Snap Inc.*, September 20, Available at: https://www.snap.com/es/news/post/the-liquid-self/ Accessed at 28.1.2019.

Kamleitner, M. 2018. 'The Psycology of Snapchat Marketing', in *Swat.io*. Available at: https://blog.swat.io/wp-content/uploads/sites/11/2016/06/Swat.io-The-Psychology-of-Snapchat-Marketing.pdf Accessed at 28.1.2019.

Kofoed, M. and Larsen, M. C. H. 2016. 'A Snap of Intimacy: Photo-Sharing among Young People on Social Media', in *First Monday*, vol. 21, Available at: https://firstmonday.org/article/view/6905/5648 Accessed at 28.1.2019.

Kryssanov, V., Tamaki, H. and Kitamura, S. 2005. 'Evolutionary Design: Philosophy, Theory and Application Tactics', in *CIRP Journal of Manufacturing Systems*, vol. 34. Available at: http://arxiv.org/pdf/es/0606039.pdf Accessed at 1.1.2019.

Lacort, J. 2018. 'Snapchat: crónica de una muerte intrascendente', in *Tecnoexplora*, July 13. Available at: https://www.lasexta.com/tecnologia-tecnoxplora/apps/snapchat-cronica-muerteintrascendente_201807125b4824740cf2cde30 6ab7283.html Accessed at 28.1.2019.

Lijenström, H. and Svedin, U. (eds.) 2005. *Micro, Meso, Macro: Addressing Complex Systems Couplings*, London: World Scientific.

Mejía, J. C. 2017. *Snapchat VS Instagram Stories: Completa comparación con diferencias y similitudes*, January 16th, 2017. Available at: https://www.juancmejia.com/redes-sociales/snapchat-vs-instagram-stories-completa-comparacion-con-diferencias-y-similitudes-infografia/ Accessed at 29.1.2019.

Niiniluoto, I. 1993. 'The Aim and the Structure of Applied Research', in *Erkenntnis*, vol. 38/1, pp. 1–21.

Obake, D. and Ito, M. 2006. 'Everyday Context of Camera Pone Use: Steps toward Technosocial Ethnographic Frameworks', in Höflich, J. R. and Hartmann, M. (eds.), *Mobile Communication in Everyday Life: An Ethnographic View*, Berlin: Franks and Timme, pp. 79–102.

Ortiz, S. J. 2015. 'Snapchat ahora permite descubrir noticias de ESPN, MTV, CNN, Yahoo …', in *IPadízate*, January 29, Available at: https://www.ipadizate. es/2015/01/29/snapchat-discover-noticias-espn-mtv-135328 Accessed at 1.2.2019.

Redacción Canela PR. 2017. 'Snapchat ha llegado para revolucionar la comunicación *online*', in *CanelaPR*, December 12. Available at: http://canelapr.com/es/snapchat-ha-llegado-para-revolucionar-la-comunicacion-online/ Accessed at 1.2.2019.

Redacción Kanlli. 2018. 'Qué es el contenido efímero y cuáles son sus ventajas para tu estrategia de contenidos', in *Innovación y nuevas ideas de Kanlli*, February 25. Available at: https://www.kanlli.com/social-media-marketing/contenido-efimero-ventajas-estrategia-contenidos/ Accessed at 1.2.2019.

Redacción Tecnosfera. 2017. 'Snap no es solo una red social: Evan Spiegel', in *El Tiempo*, October 14. Available at: https://www.eltiempo.com/tecnosfera/apps/declaraciones-del-ceo-de-snapchat-sobre-privacidad-y-facebook-232224 Accessed at 1.2.2019.

Redacción Tecnosfera. 2018. 'Por primera vez en su historia Snapchat pierde usuarios', in *El Tiempo*, August 8. Available at: https://www.eltiempo.com/tecnosfera/novedades-tecnologia/numero-de-usuarios-de-snapchat-en-agosto-de-2018-253248 Accessed at 29.1.2019.

Redacción Tecnosfera. 2019. 'Acciones de Snap siguen golpeadas tras la renuncia de jefe financiero', in *El Tiempo*, January 17. Available at: https://www.eltiempo.com/tecnosfera/apps/acciones-de-snapchat-siguen-bajando-despues-de-salida-de-jefe-financiero-315552 Accessed at 29.1.2019.

Redacción Vida Actual. 2018. 'Snapchat escuchó a sus usuarios y vuelve a su versión original', in *El País*, May 16. Available at: https://www.elpais.com.uy/vida-actual/snapchat-escucho-usuarios-vuelve-version-original.html Accessed at 29.1.2019.

Rescher, N. 1988. *Complexity: A Philosophical Overview*, New Brunswick, NJ: Transaction Publishers.

Rescher, N. 1995. *Process Metaphysics*, Albany, NY: SUNY Press.

Rubio-Romero, J. and Perlado, M. 2017. 'Snapchat o el impacto del contenido efímero', in *Telos*, vol. 107, pp. 82–92.

Salaverria, R. 2019. 'Periodismo digital: 25 años de investigación', in *El profesional de la Información*, vol. 28, pp. 1–27. Available at: https://revista. profesionaldelainformacion.com/index.php/EPI/article/view/69729 Accessed at 26.1.2019.

Scolari, C. A. (ed.) 2018. *Adolescentes, medios de comunicación y culturas colaborativas. Aprovechando las competencias transmedia de los jóvenes en el aula*, Universitat Pompeu Fabra, Barelona, Available at: http://transmedialiteracy. upf.edu/sites/default/files/files/TL_Teens_es.pdf Accessed at 16.2.2019.

Simon, H. A. 1996. *The Sciences of the Artificial*, 3rd ed., Cambridge, MA: The MIT Press.

Simon, H. A. 2002. 'Near Decomposability and the Speed of Evolution', in *Industrial and Corporate Change*, vol. 11, pp. 587–599.

Skierkowski, D. and Wood, R. M. 2012. 'To Text or Not to Text? The Importance of Text Messaging among College-Aged Youth', in *Computers in Human Behaviour*, vol. 28, pp. 744–756.

Soto, M. S. 2017. 'Evan Spiegel, el nuevo genio de Internet', in *El Tiempo*, October 14. Available at: https://proy.eltiempo.com/tecnosfera/apps/perfil-de-evan-spiegel-cofundador-de-snapchat-141252 Accessed at 1.2.2019.

Sulbaran, P. 2016. '3 cosas que han hecho de Snapchat un fenómeno entre los jóvenes', in *BBC Mundo*, April 8. Available at: https://www.ecuavisa.com/articulo/bbc/tecnologia/143406-3-cosas-que-han-hecho-snapchat-fenomeno-entre-jovenes Accessed at 28.2.2019.

The Cocktail Analysis y Arena Media. 2016. 'VIII ola Observatorio en Redes Sociales, December 13. Available at: http://www.arenamedia.com/es/blog/arena-media-y-the-cocktail-analysis-presentan-el-viii-observatorio-de-rrss/ Accessed at 1.2.2019.

Tiropanis, T., Hall, W., Crowcroft, J., Contractor, N. and Tassiulas, L. 2015. 'Network Science, Web Science, and Internet Science', in *Communications of ACM*, vol. 58, pp. 76–82.

Vega, W. 2017. 'Spectacles: lecciones de un fracaso', in *El Tiempo*, October 30. Available at: https://www.eltiempo.com/tecnosfera/novedades-tecnologia/columna-desde-la-tecnosfera-sobre-las-gafas-spectacles-146418 Accessed at 29.1.2019.

Verstraete, G., 2016. 'It's about Time. Disappearing Images and Stories in Snapchat', *Image & Narrative*, vol. 17, p. 10.

Index of Names

Page numbers followed by n indicate notes.

Subject Index

Page numbers followed by n indicate notes.

Printed in the United States
by Baker & Taylor Publisher Services